Design for Testability, Debug and Reliability

Sebastian Huhn • Rolf Drechsler

Design for Testability, Debug and Reliability

Next Generation Measures Using Formal Techniques

Sebastian Huhn
University of Bremen and DFKI GmbH
Bremen, Germany

Rolf Drechsler
University of Bremen and DFKI GmbH
Bremen, Germany

ISBN 978-3-030-69208-7 ISBN 978-3-030-69209-4 (eBook)
https://doi.org/10.1007/978-3-030-69209-4

This Springer imprint is published by the registered company Springer Nature Switzerland AG
The registered company address is: Gewerbestrasse 11, 6330 Cham, Switzerland

Preface

Several improvements in the electronic design automation flow enabled the design of highly complex integrated circuits. This complexity has been introduced to address the challenging intended application scenarios, for instance, in automotive systems, which typically require several heterogeneous functions to be jointly implemented on-chip at once. On the one hand, the complexity scales with the transistor count and, on the other hand, further non-functional aspects have to be considered, which leads to new demanding tasks during the state-of-the-art IC design and test. Thus, new measures are required to achieve the required level of testability, debug, and reliability of the resulting circuit.

This book proposes several novel approaches to, in the end, pave the way for the next generation of integrated circuits, which can be successfully and reliably integrated even in safety-critical applications. In particular, this book combines formal techniques—like the *Boolean Satisfiability* (SAT) problem and the bounded model checking—to address the arising challenges concerning the increase in *Test Data Volume* (TDV) as well as *Test Application Time* (TAT) and the required reliability.

One important part of this book concerns the introduction of *Test Vector Transmitting using enhanced compression-based TAP controllers* (VecTHOR). VecTHOR proposes a newly designed compression architecture, which combines a codeword-based compression, a dynamically configurable dictionary and a run-length encoding scheme. VecTHOR fulfills a lightweight character and is seamlessly integrated within an IEEE 1149.1 test access port controller. VecTHOR achieves a significant reduction of the TDV and the TAT by 50%, which directly reduces the resulting test costs.

Another aspect of this book concerns the design and implementation of a retargeting framework to process existing test data off-chip once prior to the transfer without the need for an expensive test regeneration. Different techniques have been implemented to provide choosable trade-offs between the resulting TDV as well as the TAT and the required run-time of the retargeting process. These techniques include a fast heuristic approach and a formal optimization SAT-based method by invoking multiple objective functions. Besides this, one contribution concerns the development of a hybrid embedded compression architecture, which is specifically designed for low-pin count test in the field of safety-critical systems enforcing a zero defect policy. This hybrid compression has been realized in close industrial cooperation with Infineon Germany. This approach allows for reducing the resulting test time by a factor of approximately three.

Furthermore, this book presents a new methodology to significantly enhance the robustness of sequential circuits against transient faults while neither introducing a large hardware overhead nor measurably impacting the latency of the circuit. Application-specific knowledge is obtained by applying SAT-based techniques as well as bounded model checking to achieve this, which yields the synthesis of a highly efficient fault detection mechanism. The proposed techniques are presented in detail and evaluated extensively by considering industrial representative candidates, which clearly demonstrated the proposed approaches' efficacy.

Bremen, Germany Sebastian Huhn

Bremen, Germany Rolf Drechsler
December 2020

Acknowledgments

First, we would like to thank the members of the research group for *Computer Architecture* (AGRA) at the University of Bremen as well as the members of the research department for *Cyber-Physical Systems* (CPS) of the *German Research Center for Artificial Intelligence* (DFKI) at Bremen. We appreciate the great atmosphere and the stimulating environment. Furthermore, we would like to thank all co-authors of the papers, which formed the starting point for this book: Prof. Krishnendu Chakrabarty, Ph.D., Dr. Stefan Frehse, Prof. Dr. Daniel Große, and Prof. Dr. Robert Wille. We also thank Infineon Germany and, particularly, Dr. Daniel Tille for a productive industrial cooperation and Dr. Matthias Sauer for various inspiring conversations. Our special gratitude addresses Dr. Stephan Eggersglüß for the inspiring talks and the long-lasting scientific cooperation. Finally, we would like to thank Pradheepa Vijay, Brian Halm, Zoe Kennedy, and Charles Glaser from Springer. All this would not have been possible without their steady support.

Bremen, Germany Sebastian Huhn

Bremen, Germany Rolf Drechsler

Contents

Acronyms

ATE	Automatic Test Equipment
ATPG	Automatic Test Pattern Generation
BCP	Boolean Constraint Propagation
BDD	Binary Decision Diagram
BMC	Bounded Model Checking
BSC	Boundary Scan Cell
BST	Boundary Scan Test
CDCL	Conflict-Driven Clause Learning
CDW	Compressed Data Word
CNF	Conjunctive Normal Form
CuT	Circuit under Test
DDU	Dynamic Decompressing Unit
DFD	Design for Debug and Diagnosis
DFR	Design for Reliability
DFT	Design for Testability
DuT	Device Under Test
EDA	Electronic Design Automation
EDT	Embedded Deterministic Test
FDM	Fault Detection Mechanism
FF	Flip-Flop
FSM	Finite State Machine
IC	Integrated Circuit
IG	Implication Graph
IJTAG	Internal Joint Test Action Group
IR	Instruction Register
JTAG	Joint Test Action Group
LPCT	Low-Pin Count Test
MCL	Maximum Codeword Length
MDL	Maximum Dataword Length
MO	Memory-Out
PB	Pseudo-Boolean

PBC	Pseudo-Boolean Constraint
PBO	Pseudo-Boolean Optimization
PI	Primary Input
PO	Primary Output
RTL	Register-Transfer Level
RTP	Rejected Test Pattern
RTPG	Random Test Pattern Generation
SAT	Boolean Satisfiability
sat	satisfiable
SBI	Single Bit Injection
SFF	Scan Flip-Flops
SoC	System-on-Chip
TAP	Test Access Port
TAT	Test Application Time
TDI	Test Data In
TDR	Test Data Register
TDV	Test Data Volume
TMS	Test Mode Select
TO	Time-Out
UDW	Uncompressed Data Word
unsat	unsatisfiable
VecTHOR	Test Vector Transmitting using enhanced compression-based TAP compression-based TAP controllers
X-value	Don't Care value

Symbols

$\mathcal{A}$	Activator signal
β	Benefit of a replacement
b_i	Bit in test data
$\bullet$	Boolean AND operator
$+$	Boolean OR operator
$\odot$	Resolution operator
$\mathcal{B}$	Set of Boolean values {0,1}
ν	Boolean variable
$\oplus$	Boolean XOR operator
ω_L	Conflict clause
$\Phi_{\#CDW}$	CNF; dictionary constraint clauses
cdw_i	Compressed data word
c_i	Chunk of compressed data word
C	(Sequiental) circuit
ω	Clause
Φ	CNF; set of clauses
Φ_{uC}	CNF; coverage completeness clauses
$\mathcal{E}$	Equality comparator signal
$\mathcal{K}$	Conflict
$\mathcal{C}$	Embedded dictionary configuration
S^*	Complete state space of sequential circuit
$\mathcal{D}$	Compressed test data
$S_D@l$	State of dictionary at time l
S_D	State of dictionary
$\varnothing$	Empty codeword
$\mathcal{Z}$	Datas container of EPs
EP	Equivalence Property
$\widehat{FF}$	Flip-flop with single transient fault
$\mathcal{F}$	Fault signal
G	Set of gates
L	Total number of hierarchical level of sequential circuit

Ω	Input test vector
IN	Set of inputs
σ	Lookup function for codewords
Ψ	Mapping function
r_s	Maximum number of reconfiguartions
Φ_{ME}	CNF; mutual exclusion clauses
$\neg X$; $\overline{X}$	Boolean negation operator
t_i	Next point in time
t_i	Current point in time
$\mathcal{N}_{det}$	Number of detected faults of $\mathcal{N}$
$\mathcal{N}$	Size of fault list
N	Set of non-robust flip-flops of sequential circuit
T	Number of scan chains
$\widehat{\mathcal{N}}$	Number of testable faults of $\mathcal{N}$
$\mathcal{O}$	Optimization function
OUT	Set of outputs
p_s	Maximum partition size
P	(Generic) partition
$\mathcal{P}_{RAND}$	Greedy-like partition enumerator
γ	Replacement of dataword
Φ_{RET}	CNF; retargeting clauses
$\overline{\Omega}$	Retargeted test vector
$\mathcal{R}$	Robustness of sequential circuit
$\widehat{S}$	Reachables states of sequential circuit
$\mathcal{P}_{SAT}$	SAT-based partition enumerator
γ	Single Bit Injection
SE	Set of sequential elements
$\mathcal{M}$	Test pattern metric for quality assessment of partition
udw_i	Uncompressed data word
$\mathcal{I}$	Uncompressed (incoming) test data
W	Set of wires

List of Figures

List of Tables

List of Algorithms

Chapter 1
Introduction

For several years, the design and fabrication of *Integrated Circuits* (ICs) no longer aim at producing devices, which fulfill one dedicated task. Instead, highly complex application scenarios are targeted, which require several heterogeneous functions to be jointly implemented on-chip at once. For this purpose, *System-on-Chip* (SoC) designs have been successfully designed, which hold several nested modules, which inevitably lead to increasing complexity in the sense of transistor count. One important step towards this is the on-going reduction of the feature size of the used technology node, which implies that a single transistor is heavily shrunk.

In fact, Gordon E. Moore postulated Moore's Law [Moo65] back in 1965. This law already predicts that the number of transistors within an IC will be doubled each 18 months, which inherently means that the integration density on the transistor-level has to be steadily increased. Compared to early microprocessors like the Intel 4004 (1971) [Cor11] holding 2,250 transistors using $10\mu m$ nodes, modern high-end microprocessors like the AMD Zen 2 Epyc Rome (2019) [Muj19] hold more than 39.5 billions of transistors with 7nm nodes, which validates the predicted exponential growth of the transistor count.

Several improvements in the EDA flow enabled the design of highly complex ICs. This complexity has been introduced to address the challenging intended application scenarios, i.e., the provided functionality and further non-functional requirements like the available computing power or the resulting power profile. On the one hand, the complexity scales with the transistor count and, on the other hand, further aspects have to be taken into consideration, which lead to new demanding tasks during the state-of-the-art IC design and test. These tasks are discussed in the next paragraph. Even though the transistor count is increasing exponentially, the die size remains nearly constant. Consequently, the size of an individual transistor—the feature size with respect to the used technological node—has to be heavily shrunk. Reconsider the above-given comparison between the Intel 4004 and the AMD Zen 2 Epyc Rome, which demonstrates a shrinking factor of more than 1000x.

S. Huhn, R. Drechsler, *Design for Testability, Debug and Reliability*,
https://doi.org/10.1007/978-3-030-69209-4_1

The designability of those highly complex systems led to the development of complex SoC designs, which can fulfill several demanding tasks at once. These systems are also introduced in safety-critical environments, for instance, in the automotive sector for implementing components like advanced driver assistant systems. One of the most important steps, besides the design and manufacturing of these systems, concerns the test of the produced devices. The circuit test is becoming even more relevant since the overall complexity of the design increases together with the complexity of the manufacturing process itself. Consequently, performing a manufacturing test of each device forms an essential step to ensure that no defects have occurred during the manufacturing. Such a defect leads to an incorrect functional behavior and, hence, possibly leads to disastrous consequences in the field of safety-critical systems or, even within non-critical applications, to customer returns and a loss of reputation.

High quality test sets are generated by the ATPG tools [Rot66, Lar92, Dre+09, EWD13], which operate on different fault models like the stuck-at fault model [Eld59, GNR64]. A fault model reflects an abstraction from the physical defects, which cannot be modeled in an accurate way with reasonable effort. One necessary prerequisite to generate these tests is about the prevailing testability of the design. Different *Design for Testability* (DFT) measures are introduced into the design to ensure a high testability. The increasing design complexity yields a significant increase in the *Test Data Volume* (TDV) and, hence, the *Test Application Time* (TAT). This increase is even more critical when testing safety-critical applications like automotive systems, which enforce a zero defect policy.

Besides the quality-oriented aspects, one crucial aspect concerns the resulting costs of the later device, which have to remain within certain margins to meet the market demand. A share about 30% of the overall costs is held by the test costs [BA13], which is even more when addressing a zero defect policy. These test costs directly scale with the different factors like the required test time per device or the number of devices that can be tested in parallel. The TDV heavily influences both aspects. Different test compression techniques like [Raj+04] have been developed aiming at a significant reduction of the TDV during the high volume manufacturing test, i.e., during the wafer test. However, these techniques are not applicable during applications like the *Low-Pin Count Test* (LPCT). Consequently, the TDV as well as the TAT is quite high during these tests since the regular test compression is not applicable. The TAT results in high test costs and—under certain conditions—the TDV may exceed the overall memory resources of the test equipment and, hence, the test is not applicable at all, which reduces the test coverage. Such a test coverage reduction harms the zero defect policy in the field of automotive testing and, hence, has to be strictly avoided. This zero defect policy requires to conduct, among others, burn-in tests. Due to the constraining environment during such a burn-in test, where only a limited number of pins is accessible and a high ratio of parallelization is required to meet the test cost margins, this is one type of LPCT.

The presented book heavily invokes solving techniques in the field of the SAT, which asks the question whether a satisfying solution for a Boolean formula exists. The SAT problem is the first problem which was proven to be $\mathcal{NP}$-complete by Cook back in 1971 [Coo71]. Since the last 30 years, a lot of research work has been spent on the development of SAT solvers, which allow to determine a satisfying solution or prove that no one exists. Even though the underlying problem is $\mathcal{NP}$-complete, the available SAT solvers are mostly able to solve the instances in reasonable run-time. A powerful extension of SAT concerns the introduction of a *Pseudo-Boolean Optimization* (PBO), which allows not just to identify a satisfying solution but an optimal one (with respect to a given optimization function and selected optimization objectives). These SAT-based techniques have been adopted to realize the *Bounded Model Checking* (BMC) [Bie+99b], which is a technique initially designed for functional verification of digital circuits by proving or disproving certain temporal properties. In this book, BMC is adopted to analyze the state space of an arbitrary sequential circuit to determine states, in which derived Equivalence Properties hold that follow a newly developed concept.

This book combines these formal techniques to address the arising challenges concerning the increase in TDV as well as TAT and the required reliability. More precisely, this book focuses on the development of VecTHOR. VecTHOR proposed a newly designed compression architecture, which combines a codeword-based compression, a dynamically configurable dictionary and a run-length encoding scheme. VecTHOR fulfills a lightweight character and is seamlessly integrated within an IEEE 1149.1 TAP controller and meant to be extending it. Such a TAP controller exists already in state-of-the-art designs and, hence, a significant reduction of the TDV and the TAT by 50% can be achieved by extending the regular TAP controller with VecTHOR, which directly reduces the resulting test costs. Furthermore, a complete retargeting framework is developed, which is meant to retarget existing test data off-chip once prior-to the transfer.

The proposed retargeting framework allows processing arbitrary test data and, hence, this can be seamlessly coupled with commercial test generation flows without the need for regenerating existing test patterns. Different techniques have been implemented to provide choosable trade-offs between the resulting the TDV, the TAT and the required run-time of the retargeting process. These techniques include a fast heuristic approach and a formal optimization SAT-based method by invoking multiple objective functions. To address even large set of test data with the formal technique, which is known to be computing-intensive, a new partition approach is developed, which incorporates the current state of the embedded dictionary and, furthermore, applies an objective function to determine the need of time-consuming configuration cycles.

A common procedure to meet the further demanding non-functional requirements concerning the computing power is about boosting the clock frequency. This approach has been seen already, for instance, in the central processing unit market for more than one decade, yielding a frequency of 4+ GHz. However, a higher clock frequency leads to a shrunk manufacturing process window. Besides this, the resulting power profile, i.e., the power consumption during different operations, is getting more and more important since new fields of application have been recently involved, where strict power constraints prevail. Examples of these applications are mobile or Internet-of-Things devices. Decreasing the transistor operation voltage is one way to meet these power constraints, which further increases the vulnerability since the voltage margin of a single transistor is reduced. Typically, these measures induce side effects like a higher vulnerability against transient faults, which occur under certain environmental conditions like high-energy radiation or electrical noise. Existing state-of-the-art measures to protect circuits like the Triple Modular Redundancy [BW89, SB89], which, however, introduces a large area overhead of more than 3x or other approaches like Razor [Ern+03, Bla+08], which heavily influences the worst-case latency of the circuit. Another important aspect concerns the resulting overall costs of the device, which is basically a contradictory objective between low costs, reliable designing as well as high quality testing.

The research focus of Chap. 8 of this work develops a new methodology to significantly enhance the robustness of sequential circuits against transient faults while neither introducing a large hardware overhead nor measurably impacting the latency of the circuit. Application-specific knowledge is conducted by applying SAT-based techniques as well as BMC to achieve this, which yield the synthesis of highly efficient *Fault Detection Mechanism* (FDM).

To summarize, the state-of-the-art chip design introduces several measures like the shrinkage of feature sizes, which allow designing highly complex IC containing up to 40 billion transistors. Further techniques like increasing the clock frequency or the downscaling of transistor voltages are applied to meet the non-functional requirements. This yields the design of highly complex SoC, which are frequently used in safety-critical applications.

New LPCT have to be applied to fulfill the requirements of zero defects and to meet the cost margins. Furthermore, the new manufacturing techniques lead to both an increased vulnerability against transient faults as well as a higher defect probability during manufacturing. Thus, these challenges have to be addressed to pave the way for the next generation of IC, which can be successfully and reliable integrated even in safety-critical applications.

The techniques of this book have been implemented and thoroughly validated. A comprehensive retargeting framework has been developed, which is written in C++. The authors make these developments publicly available at

http://unihb.eu/VecTHOR

under the terms of the MIT license. This retargeting framework has been further cross-compiled to an ARMv8A Cortex-A53 microprocessor target device, which allows emulating in combination with an electrical validation using a storage oscilloscope. The hardware developments have further been prototypically synthesized to a Xilinx XCKU040-1FBVA676 field programmable gate array device.

Parts of this book have further been published in the formal proceedings of the scientific conferences [HED16b, Huh+17b, HED17, HTD19b, HTD19a, Huh+17a], in a scientific journal [Huh+19] and were published at the informal workshop proceedings [Huh+18, HED19, HED16a]. Furthermore, the techniques [HTD19b, HTD19a] have been developed and evaluated in a tight cooperation with Infineon Germany.

This book is structured in 9 chapters plus an appendix, which are briefly summarized as follows:

Part I: Preliminaries and Previous Work

- Chap. 2 gives an overall introduction to the circuit design and the different types of test including the test generation. Furthermore, the different measures for the DFT, *Design for Debug and Diagnosis* (DFD), and *Design for Reliability* (DFR) are presented.
- Chap. 3 presents the elementary background of the utilized formal techniques to keep this book self-contained. More precisely, the SAT problem is introduced and techniques (SAT solvers) are presented, which allow solving the SAT problem effectively. This chapter further introduces SAT-based ATPG and the BMC, which are both required in Chap. 8 of this book.

Part II: New Techniques for Test, Debug, and Reliability

- Chap. 4 presents VecTHOR—a newly developed embedded compression architecture for IEEE 1149.1-compliant TAP controllers, which includes a dynamically configurable embedded dictionary implementing a codeword-based compression approach. This approach has proven itself to be effective for heterogeneous parts of the test data. To further address homogeneous parts, VecTHOR has been extended by a run-length encoding technique. This chapter introduces both the required extension in hardware, which has been integrated on top of an existing IEEE 1149.1 controller, and a retargeting framework. This framework retargets existing test data off-chip once prior to the transfer, which is implemented by using a greedy-like algorithm, which allows a fast retargeting and, however, does not reveal the untapped potential of VecTHOR.
- Chap. 5 introduces an optimization SAT-based retargeting technique, which allows determining an optimal configuration of the embedded dictionary and, hence, highly enhances the resulting compression efficacy. Thus, the complete retargeting procedure is translated into a CNF representation. This representation can be effectively processed by SAT solvers since the homogeneity of the CNF allows implementing powerful conflict analysis as well as implication techniques. Furthermore, the CNF representation is extended by *Pseudo-Boolean* (PB) aspects, which allow—in combination with a PBO solver—introducing multiple objective functions. By this, an optimal set of codewords and the corresponding compressed test data are determined.
- Chap. 6 proposes a partitioning scheme for the optimization SAT-based, which considers the current state of the embedded dictionary. By this, a further speed-up about 17.5x is achieved compared to the monolithic approach of Chap. 5. Furthermore, it introduces the concept of partial reconfiguration to the *Dynamic Decompressing Unit* (DDU), which reduces the amount of configuration data measurably.
- Chap. 7 presents a hybrid embedded architecture, which extends the state of the art to specifically address the challenges in the field of LPCT with zero defective parts per million policies as given, for instance, in the automotive applications.
- Chap. 8 presents a new DFD mechanism, which allows detecting transient faults while introducing only a slight overhead in hardware. The proposed technique analyzes the state space of the design's *Flip-Flops* (FFs) by elaborated

formal techniques SAT and BMC. At the end, application-specific knowledge is determined to describe a set of FFs (partition) and corresponding states, in which all FFs assume the same output value. Furthermore, a metric is implemented, which is inspired by the SAT-based ATPG, and allows to rate individual partition qualitatively.
- Chap. 9 summarizes the contributions of this book, concludes the results, and discusses intended future works.
- Appendix A focuses on the software side of the developed retargeting framework.

Part I
Preliminaries and Previous Work

Chapter 2
Integrated Circuits

This chapter introduces the basic principles of the IC design and test. Furthermore, measurements are introduced, which ensure that the later IC design holds a high level of testability and reliability if this is required for the intended application. In particular, the abstract circuit model is presented in Sect. 2.1 and the principles of circuit test are described in Sect. 2.2. This includes the structural test in Sect. 2.2.1 and the functional test in Sect. 2.2.2. The generation of structural test data is described in Sect. 2.3, which presents the structural test generation procedure. The basic concepts of different DFT measures are introduced in Sect. 2.4, which include the scan-based design in Sect. 2.4.1, the boundary scan test in Sect. 2.4.2 and the test access mechanism *Joint Test Action Group* (JTAG) in Sect. 2.4.3. Besides this, the basic principle of the LPCT is described in Sect. 2.4.4. Finally, this chapter presents the DFD in Sect. 2.5 and DFR in Sect. 2.6.

2.1 Circuit Model

The design of ICs involves different levels of abstraction. This abstraction is strictly required to cope with the massive complexity of nowadays systems. Such a system can consist of up to 40 billion metal-oxide-semiconductor field-effect transistors as recently demonstrated in 2019 by the AMD Zen 2 microprocessor [Muj19].
The stack of these abstraction levels [WWW06, Lie06] is presented in Fig. 2.1 as follows:

Architectural level specifies fundamental properties of the later IC and the basic architecture including, for instance, the available interfaces, number of cores, power consumption, die size, and desired clock frequency.

Register-transfer level describes the data flow between registers and logical operations using specialized hardware description languages like Verilog or

S. Huhn, R. Drechsler, *Design for Testability, Debug and Reliability*,
https://doi.org/10.1007/978-3-030-69209-4_2

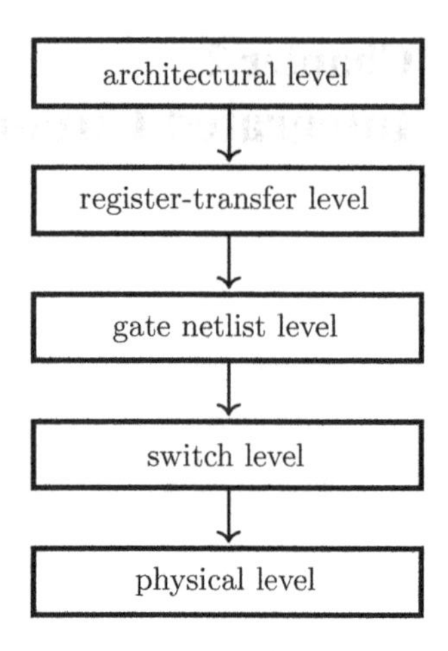

Fig. 2.1 Different abstraction level of the EDA flow

VHDL. In particular, this level defines the control signal, bus widths, word sizes, and the instruction set.

Gate level consists of logic gates and FFs, which are interconnected by wires. Typically, certain optimization techniques are applied on this level to reduce the gate count while keeping the functional behavior.

Switch level establishes a linkage between the logical and the analog level by introducing transistors, which is modeled in a discrete fashion, i.e., either switched on or off.

Physical level gives the geometric representation of the later IC design including the individual IC layers.

This book focuses on the gate-level representation, which is typically used for the conducted measurements to ensure (enhance) the testability or reliability of the resulting design, respectively. These measurements are described in Sect. 2.4 in detail.

A circuit can be modeled as a directed, acyclic graph when operating on the gate level as stated in Theorem 2.1. This graph consists of nodes and edges: A node represents a logic gate $g_i \in G$ while an edge models a wire $w \in W$. Consequently, a wire w connects two gates g_i and g_{i+1} while interpreting g_i as the predecessor of g_{i+1}. This reflects the evaluation sequence of the circuit. Thus, a gate without any predecessor is called *Primary Input* (PI) and, analogously, a gate without any successor forms a *Primary Output* (PO) of the circuit. Every gate realizes a certain function depending on its type. A wire can assume a value "0" or '1," which is then propagated/evaluated directed from the PIs towards the POs. Furthermore, a wire can also assume "Z" or "X." "Z" represents a high impedance value, which is specifically meant for bus topology with multiple (bus) driver or to isolate the output of a nested module in the IC and the *Don't Care value* (X-value) given by "X" reflects the initial state of a wire, which is not yet assigned or evaluated. The most commonly used gate types are shown in Fig. 2.2, which are reflecting the Boolean operations AND, NAND, OR, NOR, XOR, XNOR, BUF, and NOT. Furthermore,

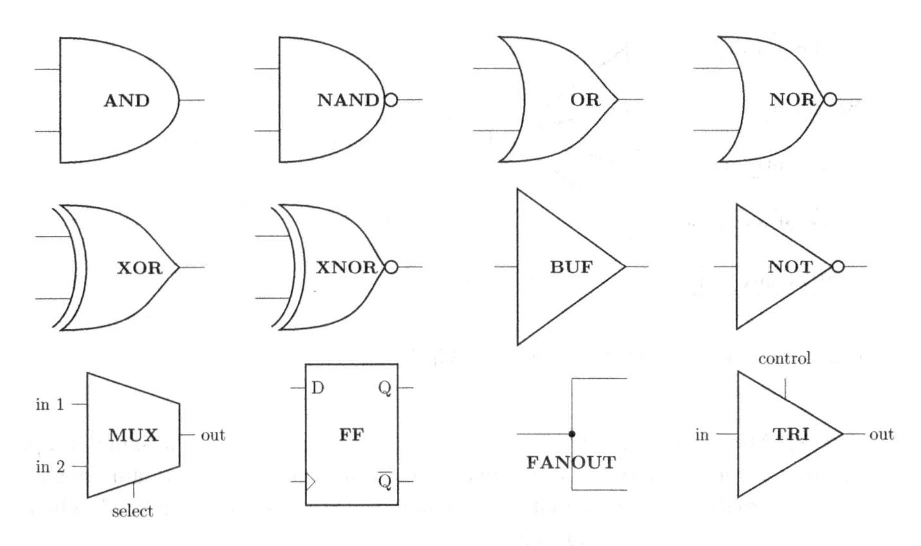

Fig. 2.2 Symbolic representation of logic gates

a FANOUT gate is introduced to represent the split of a connection consisting of the gate's input, the fan-out stem, and its outputs, the fan-out branches. Other gates are introduced to reflect a multiplexer (MUX) gate, which routes one of the two data inputs to its output depending on the select signal. Finally, a specialized TRI gate exists, which realizes a ternary logic (tri-state). This logic extends the logical values "0" and "1" by a "Z" value. Thus, the output is isolated from the input if the control signal is active. In this case, the output value is assigned to "Z."

Definition 2.1 A combinational circuit C in gate-level representation is defined as $C = (\mathsf{IN}, \mathsf{OUT}, \mathsf{G}, \mathsf{W})$ with

- IN a set of inputs,
- OUT a set of outputs,
- G a set of logic gates as introduced in Fig. 2.2 and
- W a set of wires between (gates') inputs and (gates') outputs,

which typically modeled as a directed, acyclic graph.

In contrast to a combinational circuit, a sequential circuit C contains state elements: the FFs. A FF stores a logical value "0" or "1." In general, this value is generated by the combinational fan-in cone of the FF and stored at the point in time t_i and, at t_{i+1}, fed back through its output. The points in time refer to a synchronous clock domain.[1] The state of C is defined by the values at PIs as well as the ones of the FFs. The FFs of C can be grouped by a hierarchical rank-ordered

[1]A single clock domain is assumed to ease the following descriptions. However, all proposed techniques in the remainder of this book can be extended to further clock domains.

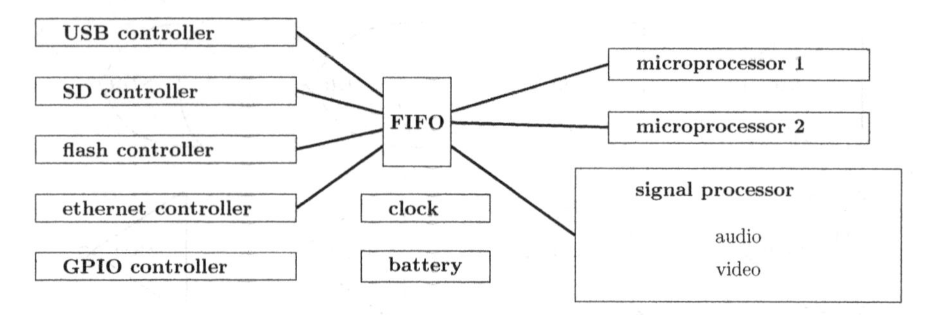

Fig. 2.3 Components of exemplary system-on-chip design

levelizing [Mic03, p. 45]. Two FFs FF_i and FF_j are contained in the same group, if the number of FFs in both fan-in cones is the same on the shortest path towards the PIs. Alternatively, a sequential circuit can also be represented by a FSM, which is described in Sect. 3.6 on page 49.

Definition 2.2 A sequential circuit C in gate-level representation extends the given Theorem 2.1 by sequential elements (FFs) and, hence, is defined as $C = (\mathsf{IN}, \mathsf{OUT}, \mathsf{G}, \mathsf{SE}, \mathsf{W})$ with

- IN a set of inputs,
- OUT a set of outputs,
- G a set of logical gates as introduced in Fig. 2.2,
- SE a set of sequential elements (FFs) and
- W a set of wires between (gates') inputs, (gates') outputs, and FFs.

2.1.1 System-on-Chip

A SoC integrates multiple components on a single substrate forming highly complex designs. The seamless integration within a SoC has major advantages compared to the naive linkage of single ICs by an interfacing circuit board. These advantages are as follows [Ben14]:

- the resulting size of the single substrate SoC is by magnitudes smaller compared to a circuit board,
- the overall power consumption is significantly lower since the integration density is higher, e.g., allows shorter wiring,
- the computing power is higher since the system runs at a higher clock frequency and, finally,
- less effort has to be spent on synchronization the single components.

Such a SoC allows addressing complex application scenarios, which require that multiple heterogeneous functions are provided. The components of an exemplary

SoC are shown in Fig. 2.3. The computation core consists of two microprocessors and a digital signal processor for conducting audio/video encoding and decoding tasks. Besides this, the design involves a wide variety of four different controllers including state-of-the-art technologies Universal Serial Bus (USB), the Secure Digital Memory Card (SD), external memory (flash) as well as an Ethernet interface. The central interface between the computation core and the controller is implemented by a first-in-first-out structure. Furthermore, the design contains a dedicated controller for the battery (here, the power management system), a clock generator, and further controlling structure for general purpose inputs and outputs (GPIOs). In fact, the SoC Fig. 2.3 describes a realistic composition of components, which is still small compared to state-of-the-art SoC design, although, it clearly demonstrates the arising challenges due to the high resulting complexity.

2.2 Circuit Test

The high complexity of the manufacturing process of ICs leads inevitably to imperfections or inaccuracies and, hence, to physical defects. For instance, such a defect is induced by process variations, surface impurities, or contact degradation. This can potentially lead to discrepancies in the later functional behavior of the manufactured IC [JG02]. This functional misbehavior may result in invalid output values, called failure. In particular for safety-critical systems, a violation of the intended functional behavior can lead to disastrous consequences if such a defective device is delivered to the customer.

The circuit test is inevitably required to prevent the risk of defects potentially harming the intended overall function of the IC. The basic circuit test principle is shown in Fig. 2.4. At first, a predetermined logic value is applied in parallel to every of the n PIs of the *Circuit under Test* (CuT), which forms a test stimulus. One depicted test stimulus is highlighted in gray in Fig. 2.4. The circuit test consists typically of a large set (here: l) of test stimuli forming the test set. As soon as one test stimulus has been processed by the circuit, the behavior of the CuT is observed at every PO in parallel, which forms, analogously to the test stimulus, the test response. During the circuit test, an individual test response is being captured for every test stimuli and, hence, l test stimuli yield l test responses. The evaluation of the circuit test is done by comparing the observed test responses with the predetermined (golden) ones proving the test result. If the observed test responses equal the predetermined ones, the test passes and, otherwise, it fails.

Mainly, two different types of tests are conducted: the structural as well as the functional test, which are both described in the next paragraphs. Note that, the basic principle as shown in Fig. 2.4 remains the same independent of the type of the applied test, although the test generation deviates significantly. Furthermore, the resulting test costs have to be taken into consideration when generating the test sets. As stated in [BA13], the test costs already hold a large share (about 30%) of the overall costs. The test costs are directly affected by, for instance, the number of test

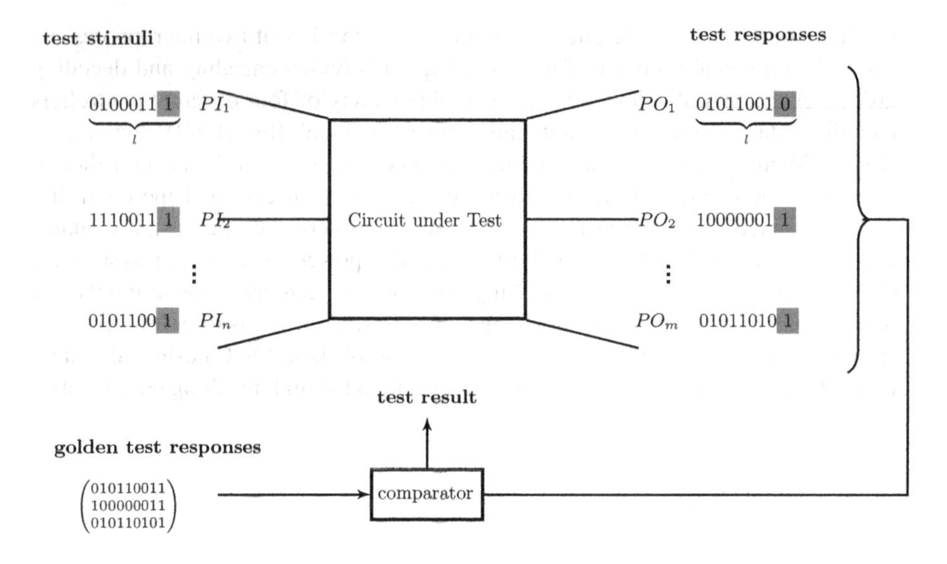

Fig. 2.4 Principle circuit test

patterns in the generated test sets and, hence, compact test sets facilitate a test cost reduction. Another important aspect concerns the overall TAT time, particularly, in an environment with limited capacities for a parallel device test like burn-in tests.

2.2.1 Structural Test

The structural test of an IC aims at proving the absence of faults, i.e., a set of test stimuli is applied to the CuT, the resulting test responses are observed and compared with the expected ones. Every test stimuli[2] is meant to detect at least one fault. It is assumed that the tested faults are not present in the circuit if the observed test responses equal the expected ones.

The concept of a fault has been introduced as a high level of abstraction of the actual defect and, in particular, their effects on the IC. For instance, such a defect is about surface impurities leading to certain physical issues during the manufacturing. Modeling these physical correlations would require a comprehensive and accurate model, which is not feasible due to

- a large number of potential defects,
- the possible—even not yet understood—interference of multiple defects and
- the large size of state-of-the-art IC designs.

[2]In the remainder of this work, a test stimulus is called test pattern.

Thus, different fault models have been published in the past, which model the effect of defects in the sense of the logical behavior of the CuT.

One of the most important fault models is the stuck-at fault model [Eld59, GNR64]. The stuck-at fault model reflects the circumstance that a wire $w \in W$ is stuck to a single logical value: A namely stuck-at-0 (s-a-0) fault at the fault site w occurs if the value of w is stuck to "0" and, analogously, a stuck-at-1 (s-a-1) fault occurs at w if the value is stuck at "1." Thus, the overall number of possible fault sites for stuck-at fault within a circuit C scales directly with the number of wires (including PIs and POs) and, hence, the overall number of stuck-at faults is $2 \cdot \#W$. Note that the fan-out branches have to be considered individually.

For proving the absence of a specific stuck-at fault at the fault site w, the opposite logic value—the fault value—has to be justified at w. Afterward, this value has to be observable at one PO of the IC and, hence, the value must be propagated from the fault site towards a PO, which enables the later comparison against the golden test response. Thus, this leads to a three-stage approach for the overall test set generation as follows:

1. Generate the list of possible stuck-at fault list for a given circuit C. Typically, different techniques are applied to reduce the number of faults significantly, for example, by taking advantage of fault collapsing with respect to the equivalence or the dominance of fault subsets [JG02, p. 79f].
   Perform for every fault the following two steps individually and append the determined test stimuli and responses to the test set:
2. Determine the necessary test stimuli to justify the opposite value at the fault site, i.e., "1" for s-a-0 and "0" for a s-a-1 fault.
3. Determine further required test stimuli to sensitize the propagation path from the fault site towards a PO such that the justified value can be observed at a PO.

Two important metrics for the structural test concern the fault and test coverage, which allow evaluating an existing test set qualitatively. The Fault Coverage (FC) is defined as follows: Given a fault list of a circuit, which contains $\mathcal{N}$ faults (with respect to a fault model), and a test set. Let the test set detect $\mathcal{N}_{det}$ faults of the fault list, then FC is calculated by $FC = \frac{\mathcal{N}_{det}}{\mathcal{N}} \cdot 100\%$. In addition to this, the Test Coverage (TC) is defined as follows: Reconsider the test set from above detecting $\mathcal{N}_{det}$ faults of a given fault list, then the test coverage is calculated by $TC = \frac{\mathcal{N}_{det}}{\widehat{\mathcal{N}}} \cdot 100\%$. Hereby, $\widehat{\mathcal{N}}$ represents the number of faults, which are detectable when considering the given circuit structure. For example, redundancies within a given circuit potentially lead to a circumstance where the fault value can never be justified without blocking the propagation path.

Example 2.3 Consider the exemplary circuit of Fig. 2.5, which consists of three gates with four PIs and one PO. The wire e is assumed as the fault site for a s-a-1 fault (Fig. 2.5a). Moreover, the opposite value, i.e., e ="0," has to be justified. Since that, the PIs of the fan-in cone have to be properly assigned with a ="1" and b ="1" as demonstrated in Fig. 2.5b. After the justification, it is required to propagate the value from fault position towards a PO to enable its observability. As

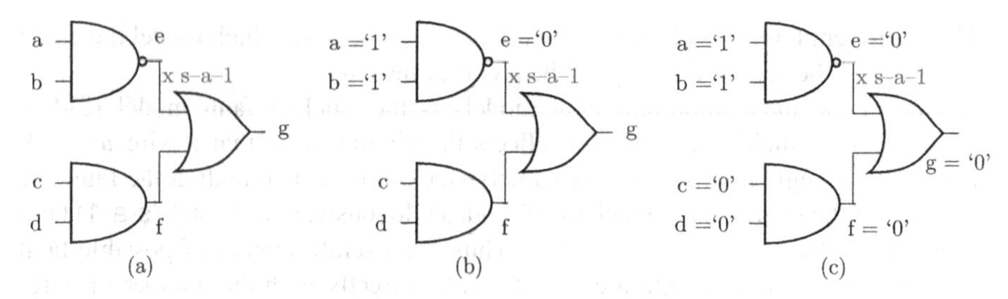

Fig. 2.5 Exemplary s-a fault. (**a**) s-a-1 at fault site e. (**b**) Justification e ="1." (**c**) Propagation

shown in Fig. 2.5c, the two further assignments c ="0" and d ="0" of the PIs are required to sensitize a propagation path.

Besides the stuck-at fault model, several other fault models exist, which aim at improving the coverage of tested faults and the possible occurred defects even more. Some of these models include several side information about the used technology nodes or the individual realization of the IC of lower abstraction levels [BA13] to model, among others, small or crosstalk delay faults [GC13, JB18]. In this book, the stuck-at fault model acts as a representative fault model candidate to conduct the structural test since it is most frequently used and, in fact, serves the baseline for every structural IC test.

2.2.2 *Functional Test*

The functional test aims at reflecting the later functional use-case. Thus, this type of test does not consider a certain fault model but actual test cases, which can be derived from the specification of the IC. It is assumed that defects lead to an incorrect functional behavior of the IC, which have occurred during the manufacturing process. These functional test cases are essential for the verification tasks [RR02] during the design phase to verify the correctness of the implementation with respect to the initial specification. Since that, these functional verification data are available and can be easily adapted for the functional test. However, an enormous search space exists, which is due to a large number of PIs. For instance, the LGA1151 is a commonly used socket to mount a 9th generation Intel central processing unit, which connects 1,151 input–output pins [Cor13]. Even if only a small share of these input–output pins, saying 100, are digital inputs, the search space grows exponentially to $2^{100} \approx 1.27 \cdot 10^{30}$, which spans the possible combinations of the PIs. Thus, the potential test coverage is quite low, even though, it can be increased by performing further functional test cases. An exhaustive test is not feasible due to the exponential growth of the search space with the number of PIs and the growth of test time and, therefore test costs. Thus, it is strictly required to incorporate the structural test, which allows determining efficient test sets.

2.3 Structural Test Generation

The test stimuli for the structural test are generated by an ATPG tool. This tool processes the netlist representation of a given IC with respect to a fault model as introduced above to generate a set of test stimuli in combination with the corresponding, expected test responses. A fault is an assumed faulty behavior of a circuit's component due to physical defects that have potentially occurred during the manufacturing process.

Two different classes of ATPG tools exist to generate the test patterns for an arbitrary sequential circuit, which either invoke well-known structural [Rot66] or SAT-based techniques [Lar92, Dre+09, EWD13]. The latter techniques are described in Sect. 3.4. Most of the structural ATPG-techniques are built on top of the D-algorithm [Rot66] or its derivatives. The D-algorithm introduces a pentavalent logic L_5 consisting of the values "0" as well as "1", the X-value, the D-value "D" and its negation "$\overline{D}$" and, hence, $L_5 = \{0, 1, X, D, \overline{D}\}$. D-values are used to represent the fault value at the fault site as soon as it has been justified and its propagation to a PO. If "D" passes a negating gate, for instance, NAND or NOR, this results in "$\overline{D}$." The structural ATPG consists of two stages as exemplarily shown in Fig. 2.5:

- The required test stimuli for the PIs are determined such that the fault value is justified at the fault site. A s-a-0 fault at w requires w ="1" as being justified, which is represented by "D." Analogously, a s-a-1 fault w requires w ="0" and is represented by "$\overline{D}$."
- The fault value "D" or "$\overline{D}$" is being propagated towards the POs on the sensitized propagation path. The gates on this path form the D-chain, whose side inputs have to be properly assigned. In the case of success, the "D" (or "$\overline{D}$") value is visible at a PO and, hence, the assumed fault is detectable.

The identification of the required side inputs—such that "D" or "$\overline{D}$" is propagated to a PI—can be derived out of the functional description of the circuit. This description depends on the actual logic gate type and is exemplarily shown for an AND gate in Table 2.1. The D-algorithm implements a two-stage approach to determine the required assignments. One major challenge concerns the case that all wires are assigned to a value; however, the faulty value has not yet been justified at the fault site or been propagated to one PO. In this case, single assignments are revised, i.e., a backtracking process is invoked, and new assignments are determined, which allow the justification or the propagation, respectively.

Table 2.1 D values of AND gate

AND	0	1	D	$\overline{D}$	X
0	0	0	0	0	0
1	0	1	D	$\overline{D}$	X
D	0	D	D	0	X
$\overline{D}$	0	$\overline{D}$	0	$\overline{D}$	X
X	0	X	X	X	X

The determination of assignments, which justify the fault value and propagate it to a PO, is a computationally hard task since the D-algorithm considers every wire as a candidate (for an assignment), which yields a large number of possible assignments. Depending on the structure of the inspected circuit, the determination of these assignments is not even possible within reasonable run-time since the D-algorithm is permanently ending up in conflicts. Different extensions of the basic D-algorithm are available to tackle these shortcomings. For instance, the Path-Oriented-DEcision-Making (PODEM) algorithm [GR81] introduces a mechanism that considers only the PIs as possible candidates for an assignment. A further extension is the FANout-oriented (FAN) algorithm [FS83] by, among others, introducing headlines in the circuit, which represents a fanout-free part of the circuit.

2.4 Design for Testability

The steadily increasing complexity of the structures, which are introduced in recent IC designs and, particularly, their integration into safety-critical applications, results in new challenges in the field of test generation. Dedicated measures are conducted to ensure a high quality of test while introducing only manageable test costs, which are typically about introducing certain DFT structures during the design phase. Depicted DFT measures are described in this section.

One elementary aspect of the DFT concerns the scan-based design, which is the state-of-the-art for large sequential circuits and described in Sect. 2.4.1. The scan-based design allows to highly reduce the complexity of the sequential test of single components. However, further challenges in the field of testing emerged with the advents of SoC designs in the 1990s. This yields the development of the *Boundary Scan Test* (BST) as described in Sect. 2.4.2. The dedicated test access mechanism JTAG [13] is described in Sect. 2.4.3. Furthermore, a highly flexible test network methodology—the *Internal Joint Test Action Group* (IJTAG) [14]—is also presented in this section, which has been recently developed to address the ever-increasing number of modules within state-of-the-art SoC designs. Besides the accessibility, another important aspect concerns the resulting test cost, which directly scales with the TDV. Finally, this section motivates the LPCT in Sect. 2.4.4.

2.4.1 Scan-Based Design

The increasing number of FFs that are included in state-of-the-art designs leads to new challenges during the test. This is due to the fact that the state space of the sequential circuit, i.e., the values of all FFs, grows exponentially with the number of FFs. Thus, a methodology is required to control and observe every single FF systematically. Otherwise, the fault coverage would decrease since the test of certain faults requires a specific state of the overall circuit, which includes the FFs.

The scan-based design [EW77] enables the controllability and observability over FFs by slight modifications of the design and the later test application as follows:

1. Regular FFs of the given sequential circuit are replaced by *Scan Flip-Flops* (SFFs).
2. SFFs are arranged sequentially by connecting the data-out (Q) pin of SFF i with the scan-data-in (SDI) pin of the SFF $i + 1$, forming the scan chain.
3. The SDI pin of the first SFF and the Q pin of the last SFF are inserted as a new PI scan-in (SI) as well as a new PO scan-out (SO).
4. A further PI is introduced to the chip-level implementing a control signal for the scan mode, the scan-enable (SE) signal. More precisely, if the SE signal is set to "1," the scan mode is active. Otherwise, the circuit remains in its functional mode, which means that the newly introduced scan structure is completely transparent.

The basic principle of a scan-based design is shown in Fig. 2.6 as follows: sA SFF is given in Fig. 2.6a consisting of a regular FF in combination with a MUX gate and the new SDI as well as SE pins. The regular FFs of an arbitrary non-scan sequential circuit are replaced by SFFs, which yield the scan-based design, which is demonstrated in Fig. 2.6b.
A test on scan-based design consists of the following three phases:

1. The initialization values for all SFFs are sequentially shifted in the circuit while the test mode is active (SE ="1").
2. The scan mode is disabled (SE="0") and the circuit operates in its functional mode for at least one clock cycle.
3. The scan mode is reactivated to shift the values of all SFFs out.

By following this scheme, the state of the circuit has been extracted after the test pattern was applied. The test result, i.e., pass or fail, is evaluated by comparing both, the state of the circuit as well as the POs values of the circuit against the predetermined one by the ATPG process. With a larger number of FFs in the non-scan design, the length of the scan chain becomes a crucial parameter. Assume a scan chain length of 50, and then the scan-in phase requires 50 clock cycles in test mode and 50 clock cycles during the shift-out phase, 100 in sum. Generally spoken, a scan chain of length k requires $2 \times k$ clock cycles for shifting and at least one further clock cycle for the functional operation. This is even more critical since the applied clock frequency is typically much lower than the one in functional mode. For reducing the number of SFFs per scan chain, multiple scan chains are introduced to the design. Hereby, it is tried to balance the lengths of all scan chains. In the example above, if two scan chains are introduced fully balanced, each of them holds just 25 SFFs and, consequently, the number of test cycles is reduced by half. In turn, each introduced scan chain requires two additional pins, the SI as well as the SO.

Due to this external accessibility of the inputs as well as outputs of the FFs, each of them is considered as a Pseudo PI and Pseudo PO, respectively. Further criteria have to be met when considering tests for a more advanced type of faults. For instance, path delay faults require two subsequent test patterns

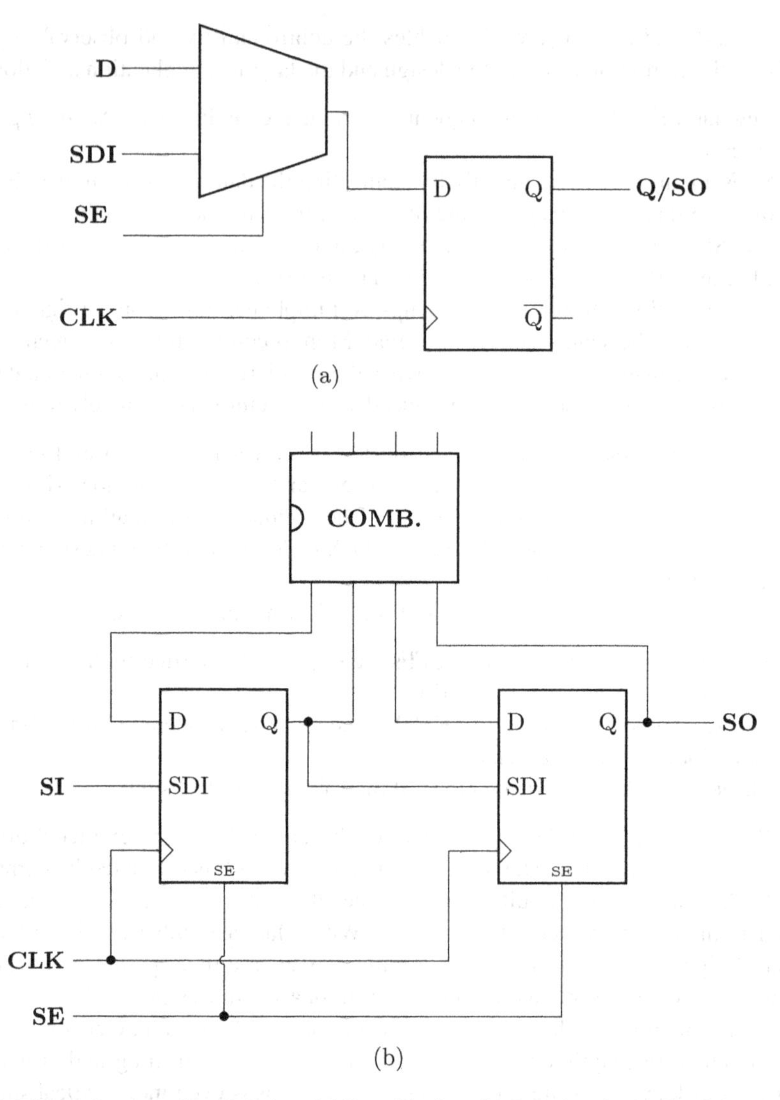

Fig. 2.6 Exemplary scan-based design. (**a**) Scan flip-flop. (**b**) Introduced scan flip-flops as scan chain

being applied to induce a transition instead of a statistic value as required when considering the stuck-at fault model. Therefore, different scan configurations are available in the literature like launch-on-capture [SP94], launch-on-shift [SP93], or enhanced-scan [DS91]. These approaches provide a trade-off between the flexibility of the second test pattern, the introduced hardware overhead, and the required computational effort for the ATPG process. In fact, the enhanced-scan gives the highest flexibility. However, this approach introduces a large overhead since the number of FFs is doubled. In contrast to this, the launch-on-capture approach implies a compute-intensive task since the ATPG process has to model the complete combinational behavior of the circuit. The launch-of-shift is more flexible since one additional shift operation is conducted, which, in turn, requires a more expensive synthesis since this operation is performed in-between two functional clock cycles.

Besides the scan test, several approaches in the field of Built-In Self-Test (BIST) have been published, which concern the Memory BIST [ZS03] as well as the Logic BIST [Het+99]. Note that the focus of this book lays on the test access level instead of the technical details of these BIST approaches.

2.4.2 Boundary Scan Test

With the increasing number of SoC designs, a new requirement arises, which concerns the controllability and observability of the boundary pins. As presented in Sect. 2.1.1, a SoC holds various components like on-chip memory modules, microprocessors or dedicated digital signal processors. When a component is embedded into a SoC then the former chip-level pins of the component become a part of the internal wiring of the SoC, which forms the boundary pins of the component. In general, the majority of these boundary pins are not accessible from the SoC chip-level pins. This is since the boundary pins are not to be meant for external access but for being interconnected with another module. However, it is strictly required for different applications to control and to observe the individual boundary pins like:

- conducting an interconnection test between two components like a microprocessor and a memory,
- testing a certain component, or
- performing certain debug operations.

Consequently, a mechanism is required, which allows the controllability and observability of every boundary pin. The BST enables access to individual boundary pins of embedded components. As presented in Fig. 2.7, a *Boundary Scan Cell* (BSC) is introduced at every boundary pin of the embedded component. Analogously to the concept of scan-based design, these BSCs are interconnected sequentially, which yields the boundary scan chain. Moreover, a boundary scan description language [13] has been developed, which allows describing the boundary scan capabilities of a component in a structured and exchangeable format. This description language enables an easy exchange of whole components between different SoC designs while retaining the existing boundary scan infrastructure. Furthermore, the structural information about the boundary scan chain is required for further steps like the test pattern generation.

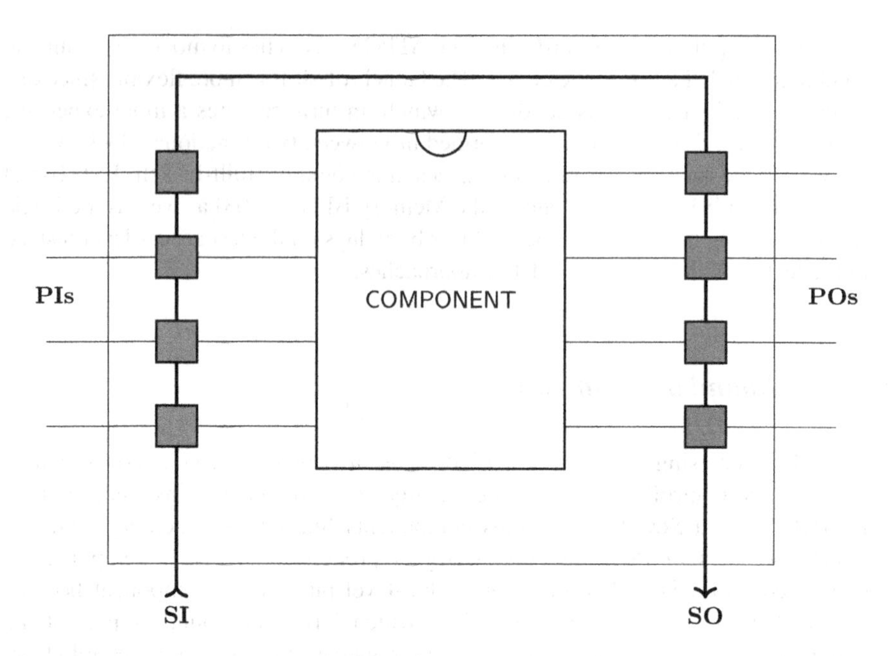

Fig. 2.7 Boundary scan chain

2.4.3 Test Access Mechanism

In particular, when considering SoC design with several nested modules, a mechanism is required to communicate with the individual component. The previous sections introduced the scan-based test to access the FFs within a single component and the BST, which implements the access to the boundary pins. Since SoC designs hold several components, all these measures have to be merged into a single mechanism. This mechanism also has to provide the interface on the chip-level pin set of the SoC and, furthermore, has to set the control signals with respect to the intended operation.

For that reason, a centralized master TAP controller is integrated into the chip-level of the SoC: Such a TAP controller provides a TAP and implements a standardized interface protocol, which typically supports different operations. These operations are selected by instruction codes, which are transferred to the TAP controller as part of the test data. When an instruction code has been received, it is decoded and executes, which controls the device.

The most frequently used mechanism is the JTAG controller. JTAG requires a TAP consisting of four (plus an optional one) pins, which have to be added to the chip-level pin set of the SoC, and is standardized within IEEE 1149.1 [13]. The four (five) signals of TAP are as follows:

- Test Clock (TCK) provides the external reduced clock for test purposes,
- Test Data-In (TDI) transfers the actual data into the device,

- Test Data-Out (TDO) allows transferring data out of the circuit,
- Test Mode Select (TMS) controls the FSM of the complete TAP controller, and
- Test Reset (TRST) is an optional signal, which implements a way to reset the TAP controller.

One major advantage of JTAG concerns the small number of input–output pins that have to be introduced at the chip-level. Every additional pin would increase both production costs and also the necessary design effort. Furthermore, it is possible to integrate multiple TAP controllers in, for instance, a hierarchical SoC design.
Besides the TAP, JTAG requires further components, which are described in the remainder of this section, as follows:

1. The TAP controller realizes the standardized protocol—supporting a set of default instructions, which can be easily extended to user-defined instructions—by implementing a FSM.
2. An *Instruction Register* (IR) that stores the recently received instruction.
3. An instruction decoder that decodes the instruction stored in the IR.
4. A boundary scan register that adapts the boundary scan chain of a component and a bypass register.
5. Further *Test Data Registers* (TDRs) implementing user-defined operations.
6. Optional registers, for instance, ones that are coupled with Logic or Memory BIST capabilities.

The complete integration of the above-mentioned components is shown in Fig. 2.8. These components are further described in the remainder of this section. Some aspects are simplified for the sake of a comprehensible visualization.

TAP Controller
The underlying FSM[3] is demonstrated in Fig. 2.9. The TAP controller implements a strict separation between loading IR data and DR data and, hence, two different branches exist within the FSM. The IR data contains an instruction sequence to control certain JTAG operations and the DR data contain the actual payload data like test data. Both branches—the IR as well as DR—contain the same states as presented in Fig. 2.9. Overall, 16 states are included, whereby six states are stable. The transition between two states is controlled by the *Test Mode Select* (TMS) signal depending on the current state of the TAP controller as well as on the actual TMS signal value. The descriptions of these states are given in Table 2.2.

JTAG Instructions
JTAG supports different operations, which, for instance, define the data sink for the next data transfer. These operations are defined by instruction codes that are transferred as a part of the data, which are typically implemented by a four-bit instruction word and, hence, 16 different instructions can be encoded. Besides the transfer of the instruction code, the TMS signal has to be properly set to enter the IR branch according to the FSM as presented in Fig. 2.9. More precisely, the instruction

[3]Note that the concept of FSM is introduced detailed in Sect. 3.6 on page 49.

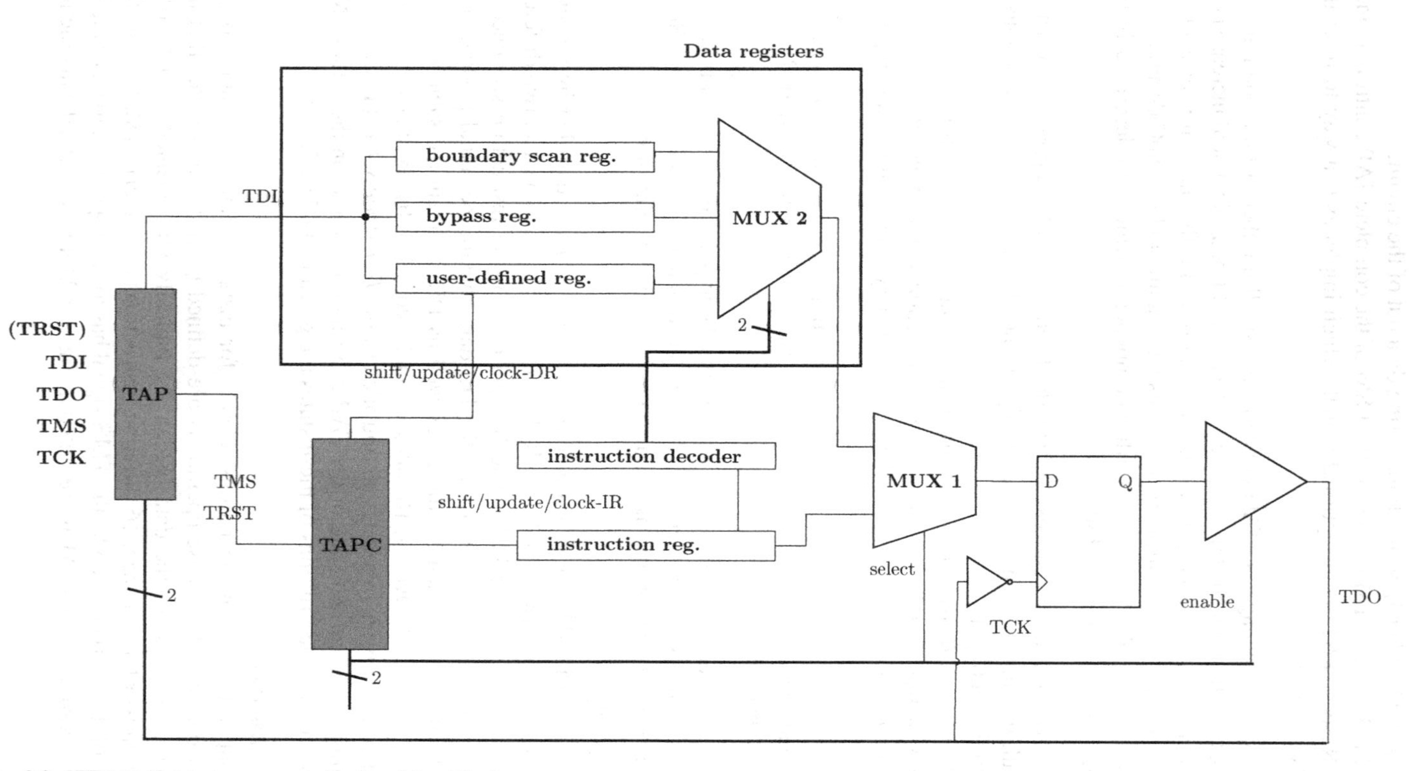

Fig. 2.8 IEEE 1149.1 test access mechanism (simplified)

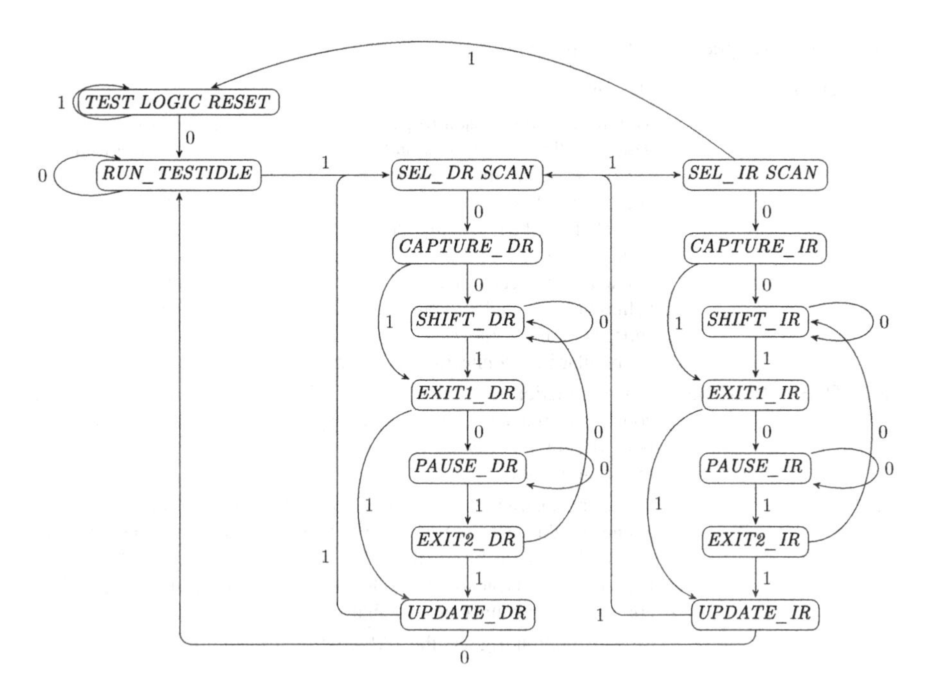

Fig. 2.9 FSM of IEEE 1149.1: TMS @ edges

Table 2.2 Description of TAP controller states

State	Description
test_logic_reset	In this state the complete test logic is disabled
run_test_idle	The test logic is only active in this state when a certain instruction has been loaded and, hence, an operation like a BIST is currently running. Otherwise, the controller is idling
select_DR_scan	This controls whether the DR branch of the FSM is to be entered
capture_DR	The associated register loads (captures) the values in parallel
shift_DR	A single bit is shifted into the corresponding register
exit1_DR	This state controls whether the pause_DR or update_DR state should be entered
pause_DR	This state allows a temporary pausing of the shifting process, for instance, to reload data off-chip
exit2_DR	This state controls whether the shift_DR or the update_DR state is entered
update_DR	In this state, the recently received instruction stored in the IR is decoded and applied

Table 2.3 Description of TAP controller instructions

Instruction	Description
EXTEST	This instruction is used to perform an interconnection test of components. At first, the corresponding boundary scan chains (registers) of the components under test are connected between the test data input and output. Afterwards, the test patterns are shifted into the boundary scan chain of the i-th component and the regular interconnections of the component i and $i+1$ are used to transfer the values into the connected boundary scan cells of the $(i + 1)$-th component. In the next step, the values of the boundary scan chain of the $(i+1)$-th component are either shifted out and compared against the expected values or transferred by the regular interconnections to a further $(i + 2)$-th component
SAMPLE/PRELOAD	This instruction is similar to the EXTEST instruction, however, the component remains in functional mode. Thus, this instruction allows both during the functional mode, the sampling of values and the injection of values
BYPASS	When the instruction is loaded, the TDI and TDO pins are directly connected through the Bypass Register and, hence, an applied value on the TDI is visible at the TDO with one clock cycle delay. This instruction is typically used to shorten the path if multiple TAP controllers are connected in a chain
IDCODE	This instruction triggers the TAP controller to emit its ID code

code is serially shifted, i.e., bit-wise, during the iteration on the shift_IR state. With the transfer of the last bit of the instruction code to the IR, the state finally traverses to the update_IR state, which decodes and executes the instruction by invoking the instruction decoder. Hereby, a set of instructions is available by default as stated in Table 2.3. Further instructions are optional, for instance, RUNBIST, which is meant to configure and start a BIST operation of a component, or INTEST, which allows performing an internal test of the component like scan-based tests. In addition to this, it is possible to implement user-defined instructions for special purposes. Typically, further TDRs are introduced in conjunction with the newly inserted instructions. As presented in Fig. 2.8, the TAP controller determines by the select control signal and a multiplexer gate (MUX 1) whether the IR or DR data are used for driving the output. The specific data register is selected by another multiplexer gate (MUX 2) with respect to the recently decoded instruction. Finally, the output signal is synchronized to the inverted TCK by a FF and controlled by a tri-state element, which drives the TDO when it is enabled by the TAP controller.

The BST is the most important operation and is still a major part of IEEE 1149.1. As previously introduced in Sect. 2.4.2, a boundary scan chain is introduced to the design, which consists of single BSCs adapting the PIs and POs of the component. The BSC can either be connected to an input or an output of the component. A BSC consists of two FFs as well as two MUX gates as presented in Fig. 2.10. Hereby, the single components have the function as follows:

- MUX 2 determines whether regular input is connected or if the path to the stored value in UP FF is established.

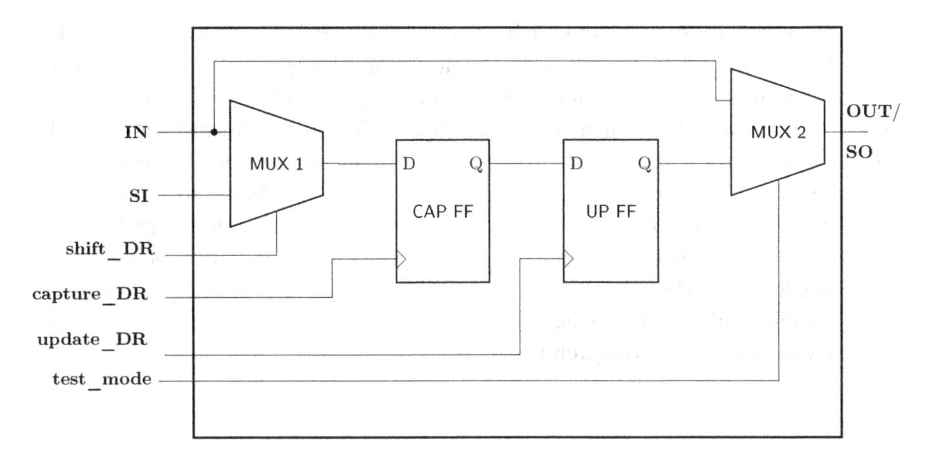

Fig. 2.10 Boundary scan cell [GMG90]

- MUX 1 decides whether the shifted or the input value gets captured.
- CAP FF is the actual register of the scan path and, hence, it allows observing the PO of the component by capturing.
- UP FF stores the value, which is applied to the component next.

Independently of the type, the test data have to be converted into a bit string since all bits are transferred strictly serialized via the TDI into the TAP controller. As a prerequisite for the transfer, the TAP controller has to be in the correct state. Reconsider that the state transitions are controlled by TMS and synchronized to TCK. Depending on the type of test, all test data may be stored in one exposed TDR, which is accessible by the CuT as well. For instance, this TDR can contain different bit fields, which represent memory addresses, control signals, or directly encoded opcodes of the CuT instruction set.

JTAG has become the de facto standard for test access in highly complex designs. However, with the increasing number of components within SoC designs, a new challenge has occurred since JTAG was not originally meant to build whole network structures for interconnecting a large number of components. Furthermore, the introduction of user-defined instructions or TDRs is rather a manual process without a strong EDA support and, hence, the process of introducing customized TDRs is time-consuming and, more importantly, prone to errors. Due to this, IJTAG has been developed, which provides a mechanism to establish the connecting test network between components while typically using JTAG TAP controllers as the top-level test access mechanism.

IJTAG

In contrast to this, IJTAG [14] provides two different standardized description languages, which are directly processed by the EDA tools to synthesize the resulting test network. The instrument connectivity language describes the access mechanism and the network, i.e., the instrument (equals the component) and the corresponding TDRs with their read/write signals. Besides this, the procedure description language describes the operations of the single instrument and, hence, allows retargeting the test data with respect to the introduced network structure. One of the most important mechanisms of IJTAG is the segment insertion bit, which determines whether a segment is on the scan path or not. By an elaborated arrangement of these elements, a large number of scan paths are (re-)configurable during the test and, hence, completely new test topologies can be implemented. However, these flexible structures have to be properly configured depending on the intended test. These configurations yield to further configuration data, which increase the resulting TDV. Due to the great potential of IJTAG, both academia and industry are focused on research in this area.

2.4.4 Low-Pin Count Test

The design of ICs includes contradictory optimization objectives, which, for instance, drive towards a high quality as well as low resulting costs. One major part of the manufacturing costs is caused by the size of the IC, which is, in turn, highly affected by the number of pins at the chip-level. These chip-level pins are very cost intensive since pads have to be introduced of which only a limited number are available without increasing the overall pad frame. As a result, the number of chip-level pins is tried to be reduced by introducing rather those pins, which are required for the later functional operation. Consequently, less pins are spent for purposes, which only concern the test or debug.

Besides introducing a LPCT to decrease the manufacturing cost, another aspect concerns the test costs: A significant less complex *Automatic Test Equipment* (ATE) can be utilized if only a LPCT is performed, which lowers the acquisition and operating costs during the test. Furthermore, LPCT in combination with a regular ATE allows to occupy multiple test sites, meaning that multiple dies (up to 128x [Nag14]) can be tested in parallel during the wafer test and, hence, the test throughput is increased. This multi-site is typically not possible when using higher test pin count set per die since the pin capacities of the ATE are limited.

2.5 Design for Debug and Diagnosis

The combination of circuit test with certain DFT measures allows achieving a high test coverage. Depending on the circuit's structure this may require performing a large number of tests. By this, it is possible to reach a high level of confidence—concerning the considered fault model(s)—that no defect has occurred during the manufacturing, which may tamper the correct functional behavior of the IC. Executing these tests yields a binary classification—in the sense of pass and fail—and, hence, the failed devices are excluded from the further production process.

Hereby, this classification does not reveal much about the reason for the failure, i.e., the actual physical defect, which has caused the failure. Even if the single failed test pattern is known, the fault (with respect to the fault model) is not yet identified since a single test pattern typically covers multiple faults. This is emphasized when applying test compaction techniques. Knowledge about this root cause is required for the overall manufacturing process, for instance, in the context of yield analysis and engineering or due to a customer return of a defective device [VG14]. This deep inspection of the IC is called silicon debug.

The yield analysis checks whether the same defects are spread over various ICs, which may indicate a systematic error. If so, the manufacturing process is slightly adjusted, which aims at preventing the initiation of these defects. This allows to enhance the yield and, hence, saves manufacturing costs, which hold a large share of the overall costs. Another aspect concerns the customer returns of devices due to malfunctions. Typically, these devices have already been operating for a longer period of time and, hence, these defects are rather due to the wear-out (during its device lifecycle) or the infant mortality than to manufacturing defects [VG02].

Analogously to the DFT, specialized DFD measures have to be introduced into the design, which allows to debug and diagnose the device in a fast and accurate fashion. These DFD measures are built on top of the DFT structures. In general, the debugging aims at sensitizing, isolating, and identifying the *bug*. This is followed by the diagnosis step, which is invoked to identify the actual root cause if possible [MV08]. The sensitization of the *bug* implies that it is replicable, which forms a prerequisite. By using the existing DFT structures, e.g., internal as well as boundary scan chains, the failing operation can be reconstructed if possible. The whole operation is typically controlled by JTAG. Another common approach is to selectively perform existing test patterns. If at least one test fails, a *new* malfunction is indicated. On the base of the failed tests, new diagnosis test patterns are generated by specialized ATPG procedures like [WLR19], which allows to further identify the defect location.

A further powerful mechanism concerns DFD structures, which enable a step-wise execution by introducing break points, for instance, if the device implements a microprocessor. The concept of breakpoint et. al. is a well-known concept from the software debugging and has been ported to a hardware mechanism, which can be seamlessly combined with JTAG. All these measurements introduce a significant overhead in the test data volume, which is even more critical since dedicated test compression techniques are not accessible in phases.

2.6 Design for Reliability

Several breakthroughs in the field of the design, fabrication, and test of ICs allowed the implementation of highly complex SoC designs. The resulting systems fulfill several missions or even safety-critical tasks at once while following a highly complex functional behavior. The enormous complexity and the intensive environmental interaction lead to a straining environment, particularly, in the case of long-term autonomous systems. Moreover, the shrinking feature size and different environmental influences, such as high-energy radiation or electrical noise, potentially cause unintended behavior. In the worst case, this leads to disastrous consequences and, hence, a high collateral damage, which has to be strictly avoided. More precisely, the shrinking feature size leads to an increased vulnerability of the FFs of a sequential circuit against single transient faults, which are typically caused by single event upsets.

Phenomena like high-energy radiation, electrical noise, particle strikes, or other environmental effects [HN06, Bau05] potentially induce a single transient fault in a vulnerable sequential circuit. Such a transient fault appears in the form of a toggled bit for a short period of time and is not logically, electrically, or temporarily masked. Thus, the output signals of the system are invalidated, which lead to an erroneous behavior. For increasing the robustness of a sequential circuit, the number of vulnerable (non-robust) FFs has to be decreased. Since that the robustness of a given circuit is an important metric, which can be derived from the number of non-robust FFs that are vulnerable to transient faults.

In order to increase the robustness of a given sequential circuit, a FDM is introduced into the circuit. This FDM is synthesized by integrating powerful mechanisms, which are capable to detect and react on occurring transient faults.

2.6.1 Robustness Assessment

A metric for robustness is required to evaluate the vulnerability of sequential circuits against transient faults. Thus, this allows measuring the fault tolerance, i.e., the robustness, with respect to a fault model [Kra+06, Doy+10]. Different robustness metrics are available depending on the considered fault model. In Chap. 8 of this book, the robustness metric of Theorem 2.4 has been used.

Definition 2.4 Let $\mathcal{C} = (\mathsf{IN}, \mathsf{OUT}, \mathsf{G}, \mathsf{SE}, \mathsf{W})$ (see Theorem 2.2 on page 12) be a sequential circuit. A FF is considered to be non-robust, if there is at least one reachable state and one transient fault such that the output behavior of C is inverted. Let N be the set of non-robust FFs with $N \subseteq \mathsf{SE}$. Then, the robustness of C can be determined as follows [Fey+11]:

$$\mathcal{R} = 1 - \frac{|N|}{|\mathsf{SE}|}$$

The non-robust FFs N can be computed by either formal methods [FD08] or by simulation-based techniques [MM10, Hua+98, NW99] to determine their robustness. When using simulation-based approaches, the robustness of a given sequential circuit C is determined as follows:

1. Define a number r of simulation cycles to be considered while adjusting the state of the circuit which is finally used for fault injection. Besides this, define a number k of cycles to be simulated for fault propagation.
2. The PIs of a given sequential circuit C are stimulated up to $r - 1$ cycles using random values.
3. In cycle r, the state s_r is extracted from the simulation environment. A copy $\widehat{s_r}$ of s_r is modified so that a single transient fault is injected at a randomly chosen FF $f \in \mathsf{SE}$ and, hence, the output value of f is toggled.
4. The circuit Φ is simulated twice for up to k cycles: One simulation starts from the healthy state s_r and one from faulty state $\widehat{s_r}$ containing the injected single transient fault. During both simulation runs and for every clock cycle, the same PIs' values are driven.
5. In cycle $r + k$, all POs of both simulation runs are compared. If at least one of the POs' values differs, the f is non-robust unless the circuit contains an FDM reporting a fault.
6. This procedure is repeated from Step 3 until all FFs are covered by the fault injection.

The number of covered states depends on the chosen parameters for r and k, which is due to the nature of the random simulation.

Chapter 3
Formal Techniques

This chapter introduces different formal techniques, which are invoked in Part II of this book, and, hence, are required for its comprehension. The described formal techniques include the SAT problem and the BMC. Furthermore, the concept of a FSM as well as a *Binary Decision Diagram* (BDD) are introduced as a symbolic model. More precisely, Sect. 3.1 introduces the Boolean algebra, defines its notation, and describes the SAT problem. Section 3.2 gives an overview of the different solving techniques, which have been developed within the last decades to address the SAT-problem effectively. Major extensions in this regard, like learning-guided approaches (see Sect. 3.2.3) or PBO driven techniques (see Sect. 3.2.4), are presented. The transformation of circuits into a formal CNF representation is introduced in Sect. 3.3. An application of SAT solving in the field of circuit testing—the SAT-based ATPG—is described in Sect. 3.4. Section 3.5 introduces the BMC, which is adopted in Chap. 8 of this book. Finally, the symbolic modeling techniques concerning FSM as well as a BDD are described in Sect. 3.6 and in Sect. 3.7, respectively, which are used in Part II of this book.

3.1 Boolean Algebra

This section briefly introduces the Boolean algebra that is the algebra of two Boolean values true and false. These values can be assumed by Boolean variables as stated in Theorem 3.1. In conjunction with logical operators, Boolean variables form Boolean expressions (see Theorem 3.2).

Definition 3.1 A Boolean variable x can assume a value from the set $\mathbb{B} \in \{0, 1\}$. By convention, the (logical) 0 is interpreted as false and the (logical) 1 as true, respectively.

S. Huhn, R. Drechsler, *Design for Testability, Debug and Reliability*,
https://doi.org/10.1007/978-3-030-69209-4_3

Definition 3.2 A Boolean expression consists of a

1. a set of n Boolean variables $X_n = \{x_1, x_2, \ldots, x_n\}$,
2. unary logical operator $\neg$ (NOT),[1]
3. binary logical operators $+$ (OR), $\bullet$ (AND), $\oplus$ (XOR), $\rightarrow$ (IMPLIES), $\leftrightarrow$ (EQUIVALENCE), and
4. parentheses.

Definition 3.3 A Boolean function f realizes a mapping of the type $\mathbb{B}^n \rightarrow \mathbb{B}^m$ with $n, m \in \mathbb{N}$. Typically, this function f is defined over a set of n Boolean variables $X_n = \{x_1, x_2, \ldots, x_n\}$.

A Boolean function f can be constructed out of any Boolean expression as stated in Theorem 3.3. For the evaluation, f has to be evaluated stepwise with respect to the set parenthesis as well as the precedences of operators. Encapsulated parts have to be evaluated first and the decreasing precedences of the operators are as follows: NOT, AND, OR, XOR, IMPLIES, and EQUIVALENCE.
The following laws are valid within the Boolean algebra:

$a \bullet b = b \bullet a$	$a + b = b + a$	commutativity
$(a \bullet b) \bullet c = a \bullet (b \bullet c)$	$(a + b) + c = a + (b + c)$	associativity
$a \bullet (b + c) = (a \bullet b) + (a \bullet c)$	$a + (b \bullet c) = (a + b) \bullet (b + c)$	distributivity
$a \bullet \overline{a} = 0$	$a + \overline{a} = 1$	complement
$a \bullet (a + b) = a$	$a + (a \bullet b) = a$	aborption
$a \bullet 0 = 0$	$a + 1 = 1$	extreme
$a \bullet 1 = a$	$a + 0 = a$	neutrality

Besides these laws, the De Morgan's law is valid, which allows to substitute all AND with OR operations, and vice versa, as follows:

$\overline{(a \bullet b)} = \overline{a} + \overline{b}$	$\overline{(a + b)} = \overline{a} \bullet \overline{b}$	De Morgan

Table 3.1 presents a truth table containing all logical operations, which are stated in Theorem 3.2. Such a truth table can be used to evaluate an arbitrary function.

[1]In this book, the negation of a Boolean variable x is written as $\overline{x}$.

Table 3.1 Logical operators

a	b	$\overline{a}$	$\overline{b}$	$a \bullet b$	$a + b$	$a \oplus b$	$a \rightarrow b$	$a \leftrightarrow b$
0	0	1	1	0	0	0	1	1
0	1	1	0	0	1	1	1	0
1	0	0	1	0	1	1	0	0
1	1	0	0	1	1	0	1	1

3.1.1 Boolean Satisfiability Problem

The SAT problem asks the question whether a satisfying solution for a given Boolean function f exists. This question is one of the most important decision problems in (theoretical) computer science and, hence, various problems have been successfully reduced to the SAT problem. The Boolean function $f : \{0, 1\}^n \rightarrow \{0, 1\}$ is classified as *satisfiable* (sat) if an assignment of all variables exists such that $f = 1$ holds. Otherwise, it is classified as *unsatisfiable* (unsat) [Bie+09].
An arbitrary Boolean function can be transformed into a normalized form, like a CNF, by applying the laws as stated in Sect. 3.1. Such a CNF Φ is a conjunction of clauses, whereby such a clause ω is a disjunction of literals and a literal represents a Boolean variable x in its positive x or negative form $\overline{x}$. The basic structure of a CNF and a satisfying assignment for an exemplary Boolean function are shown in Theorem 3.4.

Example 3.4 Let $\Phi = \underbrace{(x_1 + \overline{x}_2 + x_3)}_{\omega_1} \bullet \underbrace{(\overline{x}_1 + x_2)}_{\omega_2} \bullet \underbrace{(x_2 + \overline{x}_3)}_{\omega_3}$.
The CNF Φ consists of a conjunction ($\bullet$) of the clauses ω_1, ω_2, and ω_3. The assignments $x_1 = 1,\ x_2 = 1$, and $x_3 = 0$ satisfy Φ (sat).

This theoretical SAT problem can be used to address different research questions, which require a suitable modeling of the underlying problem to (normalized) Boolean functions. Generally, solving these functions is a hard computational task, hence, a lot of research work has been spent on developing powerful solving algorithms (sat solvers) to address this challenging problem. This is presented in the following.

3.2 SAT Solver

The SAT problem is one of the central $\mathcal{NP}$-complete problems. In fact, it is the first known $\mathcal{NP}$-complete problem that was proven by Cook back in 1971 [Coo71]. Nowadays, different SAT solvers exist, which are utilized to solve many practical problems. For instance, challenging problems in the field of EDA, like the test generation process [ED10] and the test point insertion [Egg+16a], have been reduced to the SAT problem and, hence, SAT solvers succeeded in solving them.

Algorithm 1 *DPLL*-algorithm [ES04, p.5]

```
1: while (continue) do
2:    propagate() {propagation of assignments}
3:    if !conflicts() then
4:       if hasFreeVariables() then
5:          return SAT {complete assignment without conflicts ⇒ SAT}
6:       else
7:          decide() {decision making}
8:       end if
9:    else
10:      if isResolveable() then
11:         backtrack() {backtracking conflicting assignments}
12:      else
13:         return UNSAT {non-resolvable conflict ⇒ UNSAT}
14:      end if
15:   end if
16: end while
```

In 1960, M. Davis and H. Putnam developed the DP algorithm [DP60], which allows to solve the SAT problem, i.e., to identify a satisfying solution for a given CNF if one exists or proves that no solution exists. This algorithm incorporates different logic optimization techniques based on the Boolean algebra and on an iterative application of resolution. On top of this, the DPLL algorithm was developed in 1962 [DLL62], which extends the previously proposed algorithm as follows:

1. a technique to propagate unit clauses,
2. a pure literal elimination, and
3. a backtracking-based search instead of the resolution process, which requires significantly fewer memory resources.

In fact, the DPLL algorithm still forms the basis for modern SAT solvers like Grasp [MS99], Chaff [Mos+01], BerkMin [GN02], MiniSAT [ES04], and clasp [Geb+07]. Depicted technical contributions of these solvers are discussed in detail in the remainder of this section. Note that both MiniSAT and clasp are heavily orchestrated in Part II.

Algorithm 1 presents the DPLL algorithm implementing a branch-and-bound search procedure, which processes a given SAT instance Φ to, finally, identify a satisfying solution if one exists. Such a solution consists of a set of assignments for every of the Boolean variables that are included in the given SAT instance. More precisely, the search algorithm iteratively selects a Boolean variable that is not yet assigned (said to be free) and chooses a value ("0" or "1") for its assignment. If all Boolean variables are assigned, i.e., no further decision can be made, the algorithm returns sat (lines 4 to 5). Otherwise, the algorithm selects a free Boolean variable by invoking specific heuristics (line 7). This assignment is then propagated to all clauses (line 2).

Besides these decision-driven assignments, the DPLL algorithm implements the *Boolean Constraint Propagation* (BCP), which takes advantage of a specific type of clauses: the unit clauses. A unit clause holds exactly one free Boolean variable and is not yet satisfied, as shown in Theorem 3.5. Consequently, only one possibility exists to satisfy this clause, which is about assigning the Boolean variable with respect to its polarity such that the corresponding clause becomes satisfied. Due to the structure of the CNF, i.e., every single clause has to be satisfied to satisfy the whole CNF, and an implication can be derived out of this scenario. The implied assignment is propagated to the complete SAT instance, which may lead to further unit clauses, which are then processed iteratively. Theoretically, it is possible to prove that a given SAT instance is unsat by performing the BCP without making a single heuristic decision at all, which, of course, requires a specific structure of the given SAT instance. Not least because of this, the SAT solving procedure starts with an initial BCP.
The above-mentioned steps are only executed if no conflict has occurred. A conflict occurs if a variable is assigned contradictorily to both—"0" and "1"—at the same time to satisfy two or more unit clauses. In this case, the propagation is begin interrupted, leading to the bound.

Example 3.5 Reconsider clause ω_1 of Theorem 3.4 with $\omega_1 = x_1 + \overline{x_2} + x_3$. Given the assignments $x_1 = \text{"}0\text{"}$ and $x_2 = \text{"}1\text{"}$ (while x_3 is not assigned), ω_1 is a unit clause. The assignment of x_3 can be implied with $x_3 = \text{"}1\text{"}$ since no other solution exists to satisfy the clause under the current assignment.

The resolution of these conflicts is an essential part of a SAT solving algorithm. More precisely, the assignment has to be undone to resolve the conflict, which leads to this impasse: the backtracking. The SAT solver stores certain information during the solving process, which allows backtracking specific assignments to, finally, resolve the conflict and to *flip* the last decision value. The required information is stored in a level-wise fashion forming the Decision Level and, hence, the backtrack history. In the case of the DPLL algorithm, all backtracking steps are conducted in a strict sequential order. This means that the assignments are successively undone until the conflict is resolved (line 11). If the conflict cannot be resolved, the SAT solver returns unsat (line 13).

3.2.1 Decision Heuristic

It is required to decide if a state is reached during the solving process, in which neither further implications—by performing a BCP—can be deduced nor a conflict has occurred. This decision is about a specific assignment ("0" or "1") of a Boolean variable, which is demonstrated in principle in Theorem 3.6. Several heuristics have been proposed by the literature to determine a *good* decision.

In general, the computational effort introduced by the decision heuristic is competing with the efficacy in terms of how much the solving procedure could have been accelerated. The computational effort strongly varies depending on the heuristics since the complexity varies from a purely random selection of a free Boolean variable to individually counting while considering aging aspects. Some of the most frequent decision heuristics are discussed in the following.

Example 3.6 Consider the CNF Φ with

$$\Phi = \underbrace{(\overline{x_1} + \overline{x_3} + x_4)}_{\omega_1} \bullet \underbrace{(x_1 + \overline{x_2} + x_3)}_{\omega_2} \bullet \underbrace{(x_1 + \overline{x_2} + \overline{x_3})}_{\omega_3} \bullet \underbrace{(\overline{x_1} + x_3)}_{\omega_4}$$

consisting of four clauses and four Boolean variables. Let all the Boolean variables be unassigned initially (t=0). At this point in time, no implication can be deduced since none of the clauses is a unit clause and, hence, the SAT solver invokes a decision heuristic to select a Boolean variable, including its assignment. Let x_1 = "1" be selected as the decision ($t = 0$) and be propagated to all three clauses. The positive polarity of x_1 in ω_2 and ω_3 causes that the literal is locally satisfied (as indicated in green) and, hence, both clauses are completely satisfied as well (as indicated in light gray). However, $\overline{x_1}$ is neither satisfied in ω_1 nor in ω_4 (as indicated in red).

$$\Phi = (\overline{x_1} + \overline{x_3} + x_4) \bullet (x_1 + \overline{x_2} + x_3) \bullet (x_1 + \overline{x_2} + \overline{x_3}) \bullet (\overline{x_1} + x_3)$$

Moreover, clause ω_4 becomes a unit clause yielding the implication x_3 = "1" ($t = 1$) since this assignment is required to fulfill the ω_4 and, in particular, the overall CNF Φ. After propagation of x_3, clause ω_4 gets satisfied and ω_1 becomes a unit clause.

$$\Phi = (\overline{x_1} + \overline{x_3} + x_4) \bullet (x_1 + \overline{x_2} + x_3) \bullet (x_1 + \overline{x_2} + \overline{x_3}) \bullet (\overline{x_1} + x_3)$$

A further implication x_2 = "1" is deduced ($t = 2$), which—after its propagation—fulfills ω_1 and, by this, the complete CNF Φ is satisfied.

$$\Phi = (\overline{x_1} + \overline{x_3} + x_4) \bullet (x_1 + \overline{x_2} + x_3) \bullet (x_1 + \overline{x_2} + \overline{x_3}) \bullet (\overline{x_1} + x_3)$$

The Maximum Occurrence of clauses of Minimum size (MOM) heuristic invokes a maximizing function, which incorporates the recent assignment to identify a beneficial candidate. More precisely, preference is given to satisfying small clauses and, furthermore, the number of occurrences of a Boolean variable is considered with respect to its polarity [JW90]. Besides this, BOHM [BS96] introduces the first look-ahead procedure. One frequently used decision heuristic is the Dynamic Largest Individual Sum one as introduced in Grasp [MS99], which provides a good trade-off between the efficacy and computational overhead. This heuristic determines the Boolean variable, which occurs most frequently in clauses that are not yet satisfied. The Variable State Independent Decaying Sum is another approach that is integrated into Chaff [Mos+01], which determines a score for each Boolean variable individually. This heuristic aims at combining the advantages of previously proposed ones. A counter is assigned to every Boolean variable, which is increased if this variable occurs in a not-yet-satisfied clause. All counters are adjusted periodically, e.g., by a division with a constant value, to implement an aging approach. In the case of decision making, the Boolean variable with the highest counter value is used.

In general, the direct comparison of different heuristics is complicated since the efficacy depends on the SAT instance structure and other aspects of the SAT solver concerning its learning capabilities.

3.2.2 *Restart*

Another feature concerns the restarts, which have proven themselves to be quite effective. Restarts are performed during the search procedure, for instance, in a random fashion. By this, it is avoided that the SAT solver remains in a non-solution region of the overall search space. If a restart is performed, the assignments in combination with the derived implications are deleted. However, some information is kept to ensure that the SAT solver traverses the search space differently. The frequency of performing restarts forms a crucial parameter, in particular, when processing large unsat instances. These instances typically require more run-time being spent on the exhaustive search to prove the unsat character. Thus, adaptive techniques have been recently proposed to determine the frequency of restarts, which is not counterproductive [Bie08].

3.2.3 *Conflict-Driven Clause Learning*

The Grasp algorithm [MS99] enhances the conflict resolution technique of the classical DPLL-algorithm by a *Conflict-Driven Clause Learning* (CDCL) approach. During the search procedure, the SAT solver enters potentially a region of the search space, in which no satisfying solution exists with respect to the structure of the given CNF and the recently made decisions. In such a case, a conflict inevitably occurs, which is then resolved by the classic DPLL-procedure by flipping the most recent decision of the Boolean variable or backtrack to a previous decision level. In both of these cases, the SAT solver remains in nearly the same region of the search space. Assuming that no satisfying solution exists in this local region of the search space, the classic algorithm has to explore this part exhaustively without any success. The CDCL approach addresses this shortcoming and allows preventing such a situation by conducting a conflict analysis. More precisely, the principle of CDCL allows not just to resolve the conflict but excludes the root cause of the occurred conflict during

the remainder of the search. This exclusion avoids the occurrence of prospective conflicts, which are similar—from a structural point of view—to the previously seen ones. This procedure accelerates the solving process since, in general, the search space is traversed more effectively.
Basically, the CDCL invokes the following three steps:

1. Determine the assignments forming the root cause for the occurred conflict.
2. Deduce a conflict clause of these conflicting assignments and append the newly generated clause to the original CNF instance.
3. Backtrack the search procedure to a decision level, which is derived from the recently determined conflict clause. This level is not necessarily the last chronological ordered decision level.

The CDCL extends the DPLL algorithm by introducing mechanisms to conduct

- a non-chronological backtracking technique,
- a root cause analysis of the occurred conflict, and
- a clause learning infrastructure to add and manage learned clauses to the CNF instance.

The root cause analysis of a conflict that has been recently occurred requires further information about the search process, which has been conducted so far. In particular, information about the deduced implication—including their corresponding clauses and decision levels—is required. This information is sampled during the search and stored in a certain graph-based structure: the *Implication Graph* (IG). The following two paragraphs describe the non-chronological backtracking and introduce the structure of the IG.

Non-Chronological Backtracking
The classic DPLL algorithm follows a strict chronological backtracking approach, which might lead to a state during the search, where it gets trapped in non-solution regions of the search space. In contrast to this, the non-chronological backtracking as introduced by [MS99] allows to resolve conflict effectively if the circumstance is as follows: Assume that a conflict occurs in clause ω_c in Decision Level i after a performed BCP. Further assume that the last decision, which affects ω_c has been made in decision level j with $j << i$. The resolution of this conflict requires to backtrack at least to decision level j. The DPLL algorithm does not allow to backtrack to j directly but requires a stepwise backtrack between all decision levels l with $l = i - j$, which involves a time-consuming procedure.

Implication Graph
An IG is a skew-symmetric directed graph containing the variable assignments associated with the corresponding decision level. Every vertex of the IG represents an assignment of a variable and every directed edge indicates an implication[2] with the

[2]Note that made decisions have no incident edges.

corresponding clause. The IG is constructed during the decision making as well as the BCP phase.

1. The IG is cut into two partitions such that one contains the conflict while the other one holds all decisions and implications leading to the conflict.
2. The partial assignment—formed by all related variable assignments—is extracted, which has an outgoing edge crossing the cut.
3. The partial assignment is negated and added as a newly learned conflict clause to the problem instance.
4. The backtrack level is determined, which is the highest decision level in the learned clause. This strategy has an advantage since the learned clause becomes unit (or assertive) under the resulting partial assignment [Mos+01].

From a functional point of view, a sequence of resolution operations—as stated in Theorem 3.7—is performed on clauses during the clause learning procedure [MS00], which yields a temporary clause after each performed step. The resolution operation terminates with the learned clause as the result if the current decision level contains just one remaining variable (in any generated clause). The literals of the learned clause are the reason for the conflict, namely reasoning literals. The principle of the resolution operation is demonstrated in Theorem 3.8.

Definition 3.7 Let $\odot$ represent the resolution operator. Given two clauses ω_j and ω_k, for which a unique variable x exists such that x is contained in both clauses ω_j and ω_k in opposite polarities. Then $\omega_j \odot \omega_k$ contains all the literals of ω_j and ω_k except x and $\overline{x}$.

Example 3.8 Let $\Phi = \underbrace{(x_1 + x_4 + x_5)}_{\omega_1} \bullet \underbrace{(x_4 + x_6)}_{\omega_2} \bullet \underbrace{(\overline{x_5} + \overline{x_6} + \overline{x_7})}_{\omega_3} \bullet \underbrace{(x_3 + x_{10})}_{\omega_4} \bullet \underbrace{(x_2 + x_7)}_{\omega_5}$ consists of the five clauses ω_1 to ω_5. The introduced notation to represent the decision and propagation process is as follows: $x_i = b@n$ with x_i is the variable, b the assigned logic value, and n the decision level. For example, $x_1 = 0@1$ means that the variable x_1 is assigned to the value "0" at decision level 1. The corresponding IG is shown in Fig. 3.1, which indicates that the clause ω_3 is unsatisfied since all literals evaluate to false due to

- the implication on x_7 in ω_5 at decision level 2, and
- the implications on x_5 in ω_1 and x_6 in ω_2 at decision level 4.

The resulting conflict is denoted by $\mathcal{K}$, whose analysis starts with the unsatisfied clause ω_3. First, the resolution operation is executed on ω_2 and ω_3 (which is one of the clauses at decision level 4) as follows:

$$\underbrace{(\overline{x}_5 + \overline{x}_6 + \overline{x}_7)}_{\omega_3} \odot \underbrace{(x_4 + x_6)}_{\omega_2} = \underbrace{(x_4 + \overline{x}_5 + \overline{x}_7)}_{\omega_T}$$

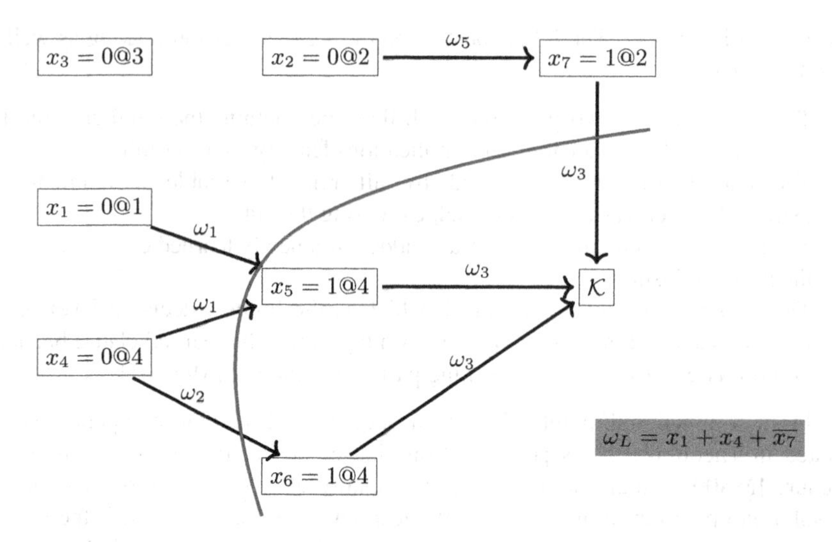

Fig. 3.1 Implication graph with occurred conflict

The temporary clause ω_T includes all literals of both clauses except x_6 since x_6 is included in opposite polarities in ω_2 and ω_3. Afterwards, the resolution operation is executed on ω_T and ω_1 as follows:

$$\underbrace{(x_4 + \overline{x}_5 + \overline{x}_7)}_{\omega_T} \odot \underbrace{(x_1 + x_4 + x_5)}_{\omega_1} = \underbrace{(x_1 + x_4 + \overline{x}_7)}_{\omega_L}$$

The clause ω_L contains all literals of both clauses except x_5 since x_5 occurs in opposite polarities in ω_T and ω_1. The clause ω_L is treated as the learned clause since it holds all reason literals and just one literal remain unprocessed at the decision level 4, which fulfills the condition to terminate. Consequently, ω_L is added to the CNF instance $\Phi' = \Phi \bullet \underbrace{(x_1 + x_4 + \overline{x_7})}_{\omega_L}$. The corresponding cut of the implication graph is shown in Fig. 3.2. More precisely, a backtracking operation is performed to the highest decision level of the literals, which are involved in the newly generated conflict clause ω_L. Here, x_4 holds the highest decision level of 4 and, hence, the backtracking removes the assignment $x_4 = 0@4$. By this, the clause ω_L becomes unit and, hence, x_4 is implied to "1." If this implication leads to a conflict, a new conflict clause is generated directly. The variables, which are then involved in this new conflict clause, determine the backtrack level and, hence, it is jumped back even further in time (as reflected by the decision level).

In the remainder of the SAT solving process, x_4 is implied to "1," decisions made on x_3 and x_5 with x_3 = "0" and x_5 = "1," which fulfill CNF instance. The final implication graph is depicted in Fig. 3.2.

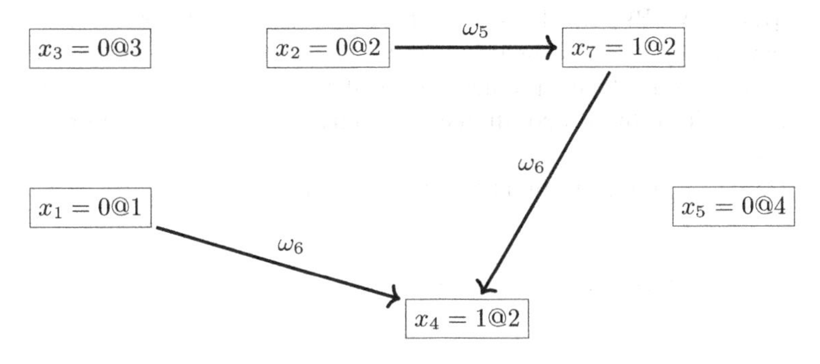

Fig. 3.2 Resulting implication graph

Conflict Clause Management
One crucial aspect in the context of clause learning concerns the management of newly generated conflict clauses. Modern SAT solvers are applied on highly complex CNF instances, which generate a vast number of conflict clauses within a concise time. This inevitably leads to a blast of the necessary memory resources. Moreover, this introduces a significant overhead during BCP since learned clauses have to be regularly processed as well. Thus, modern SAT solvers like BerkMin [GN02] invoke smart heuristics periodically to erase the depicted conflict clauses from the database. In particular, metrics determine the activity—in the sense of how often the clause was processed—of every single clause as a measurement for its relevance and, hence, less relevant clauses are erased first. This procedure ensures that the size of the conflict clause database is not exceeding a certain limit, which would introduce a bottleneck to the overall search procedure.

3.2.4 Optimization-Based SAT

In contrast to regular SAT solvers, optimization-based SAT solvers like clasp [Geb+07] enable to evaluate the quality of a determined solution. The underlying search procedure, i.e., the search for a set of assignments such that the SAT instance is satisfied, is modeled as an optimization problem. The criteria for the qualitative evaluation are given as PB co-factors and, hence, the PB-SAT problem allows for an integration of weights. The PB-SAT instance $\Phi : \{0, 1\}^n \rightarrow \{0, 1\}$ consists of conjugated constraints $\sum_{i=1}^{n-1} c_i \bullet \widehat{x_i} \geq c_n$ using $c_1, \ldots, c_n \in \mathbb{Z}$ as weights and $\widehat{x_i}$ as positive or negative literals.

Formally, the PBO extends the PB-SAT problem such that an objective function $\mathcal{O}$ can be seamlessly integrated to assess the quality of the determined solution. By this, it is not only possible to give an arbitrary solution as regular SAT solvers do but to determine the optimal solution concerning $\mathcal{O}$. PBO-based or similar optimization-based procedures have already been successfully applied in the testing domain,

e.g., in [EWD13, SBP16]. Although dedicated solving algorithms exist for this kind of problem, many of these algorithms use SAT solving techniques internally. The PBO problem is one of these problems. Here, the PB-SAT instance, i.e., the Boolean formula in CNF or the PB constraints, respectively, is extended with an objective function $\mathcal{O}$.[3]
Typically, the objective function $\mathcal{O}$ is given as a linear sum:

$$\mathcal{O}(x_1, \ldots, x_k) = \sum_{i=1}^{k} m_i \bullet \widehat{x_i} \text{ with } m_1, \ldots, m_k \in \mathbb{Z}$$

The result of $\mathcal{O}$ is the arithmetic sum of all constants m_i associated with a literal $\widehat{x_i}$, which evaluates to true under a given assignment. Basically, a PBO SAT solver utilizes the minimization as a solving target, i.e., it returns the solution that minimizes $\mathcal{O}$.

Example 3.9 Let $\Phi = (3x_1 + 4\overline{x_2} + x_3 \geq 3) \bullet (3\overline{x}_1 + 4x_2 \geq 2) \bullet (4x_2 + \overline{x}_3 \geq 4)$ and $\mathcal{O} = 1x_1 + 1x_2 + 1x_3$. In this case, the solution $x_1 = 1,\ x_2 = 1$, and $x_3 = 0$ satisfies the given PB-SAT instance and, at the same time, minimizes the outcome of the objective function $\mathcal{O}$ ($\mathcal{O} = 2$). In contrast, the solution $x_1 = 1,\ x_2 = 1$, and $x_3 = 1$ also satisfies the instance, but results in $\mathcal{O} = 3$, which is higher than in the previous solution.

Modern PBO SAT solvers also support multiple objective functions $\mathcal{O}_1, \mathcal{O}_2, \ldots, \mathcal{O}_n$. Here, priorities are used: At first, $\mathcal{O}_1$ is used as an objective function and, afterwards, the solution is improved concerning $\mathcal{O}_2, \ldots, \mathcal{O}_n$. However, it is not possible to decrease the result of the objective function with a higher priority.

3.3 Circuit-to-CNF Transformation

The application of SAT solvers, e.g., to address challenges in the field of digital circuit design, implies that the circuit has to be transformed into a representation, which is compatible with a SAT solver. Thus, a CNF has to be generated that reflects the functional behavior of the given sequential circuit. The transformation can be conducted by following the scheme proposed in [Tse83, Lar92]. Reconsider that a combinational circuit C is defined by $C = (\mathsf{IN}, \mathsf{OUT}, \mathsf{G}, \mathsf{W})$, as stated in Theorem 2.1 on page 11.
The basic principle of the transformation is about assigning a new Boolean variable x to every wire $w \in \mathsf{W}$. In the case of a FANOUT, which is connected to the wire w, the FANOUT stem, as well as its branches, are assigned to the same

[3] Since a Boolean formula in CNF can be easily transformed into PB constraints and modern PBO solvers typically accept CNFs as input, we use the notion of CNF in this paper if possible.

Table 3.2 Minimized CNF of logic gates [Lar92]

Logic gate	Boolean operator	Minimized CNF
NOT	$A = \overline{B}$	$(\overline{B} + \overline{A}) \bullet (B + A)$
AND	$A = B \bullet C$	$(\overline{B} + \overline{C} + A) \bullet (B + \overline{A}) \bullet (C + \overline{A})$
OR	$A = B + C$	$(B + C + \overline{A}) \bullet (\overline{B} + A) \bullet (\overline{C} + A)$
NAND	$A = \overline{(B \bullet C)}$	$(\overline{B} + \overline{C} + \overline{A}) \bullet (B + A) \bullet (C + A)$
NOR	$A = \overline{(B + C)}$	$(B + C + A) \bullet (\overline{B} + \overline{A}) \bullet (\overline{C} + \overline{A})$
XOR	$A = B \oplus C$	$(\overline{B} + \overline{C} + \overline{A}) \bullet (B + C + \overline{A})$ $\bullet(B + \overline{C} + A) \bullet (\overline{B} + C + A)$
XNOR	$A = \overline{(B \oplus C)}$	$(\overline{B} + \overline{C} + A) \bullet (B + C + A)$ $\bullet(B + \overline{C} + \overline{A}) \bullet (\overline{B} + C + \overline{A})$

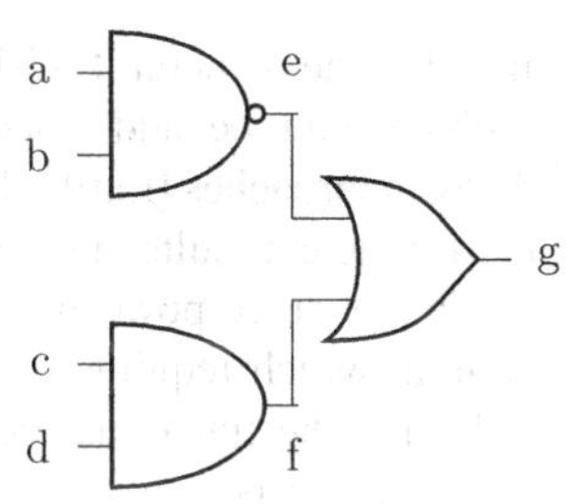

Fig. 3.3 Exemplary circuit

Boolean variable x. The logic functionality of every gate $g \in G$ is deduced and replaced with the corresponding (minimal) gate CNF Φ_g, as stated in Table 3.2. These CNFs reflect the characteristic function, i.e., the correct functional behavior of the corresponding gate type is ensured. The overall circuit CNF Φ is successively built from the conjugation of all individual gate CNFs such that $\Phi = \Phi_{g1} \bullet \ldots \bullet \Phi_{g_N}$ with $N = \#G$. An exemplary transformation is demonstrated in Theorem 3.10. One significant advantage of this gate-wise transformation concerns the resulting complexity, which is of linear size. More precisely, the number of Boolean variables in the resulting CNF equals the number of wires $w \in W$ contained in the given circuit C. Furthermore, the number of gate CNFs N scales linearly—depending on the gate type, e.g., two for a NOT gate—with the number of gates $g \in G$ and, hence, $O(N)$. However, one drawback of this transformation concerns the loss of structural knowledge about the *global* circuit since the SAT solver, e.g., the BCP, operates gate locally as well. Different approaches have been published in the literature, which aims at reintroducing these structural relationships to improve the efficacy of the CNF [Vel04].

Example 3.10 Consider the exemplary circuit given in Fig. 3.3, which is to be transformed into a CNF. Initially, the stated wires $a, b, \ldots, g$ are assumed as Boolean variables $x_a, x_b, \ldots, x_g$. The logic gates are I.) transformed into the Boolean function (with respect to the wires) and II.) substituted with the corresponding (minimal) gate CNF given in Table 3.2. Thus, the transformation is—I.) on the left and II.) on the right—as follows:

$$e = \overline{a \bullet b} \qquad \Phi_e = (\overline{a} + \overline{b} + \overline{e}) \bullet (a + e) \bullet (b + e)$$

$$f = c \bullet d \qquad \Phi_f = (\overline{c} + \overline{d} + f) \bullet (c + \overline{f}) \bullet (d + \overline{f})$$

$$g = e + f \qquad \Phi_g = (e + f + \overline{g}) \bullet (\overline{e} + g) \bullet (\overline{f} + g)$$

$$\Rightarrow g = (\overline{a \bullet b}) + (c \bullet d) \qquad \Phi = \Phi_e \bullet \Phi_f \bullet \Phi_g$$

3.4 SAT-Based Test Generation

Besides the structural ATPG algorithms as introduced in Sect. 2.3, the ATPG problem can be addressed by orchestrating SAT-based techniques. SAT-based ATPG approaches [Lar92, Dre+09, EWD13] allow determining test pattern for even hard-to-detect faults in complex designs.

This is since powerful SAT solvers are applied for determining the actual test pattern, which requires a transformation of the circuit into a CNF representation reflecting the functional behavior. The resulting CNF is then processed by a SAT solver and, especially, used to extract the test stimuli after a satisfying solution has been determined.

More precisely, every gate is successively transformed into a CNF and, afterwards, concatenated as described in the previous Sect. 3.3. The SAT-based ATPG is inspired by the Boolean Difference, which is demonstrated in Theorem 3.11. This model consists of the three components as follows:

1. the fault-free circuit,
2. the faulty circuit (starting from the fault site), and
3. a Miter consisting of a XOR gates, which are pairwise comparing the POs of the fault-free with the faulty circuit. The outputs of these XOR gates are then fed into a final OR cascade, which reflects the Boolean Difference [Mic03].

Example 3.11 Given the exemplary circuit of Fig. 3.4a, which acts as the fault-free (good) circuit G. This circuit can be represented as the Boolean function as follows:

$$G = [f = d \bullet e] \bullet [e = \overline{c}] \bullet [d = a \bullet b]$$

By applying the laws of the Boolean algebra, the Boolean function G can be transformed into the CNF as follows:

$$\Phi_G = (\overline{d} + \overline{e} + f) \bullet (d + \overline{f}) \bullet (e + \overline{f}) \bullet (\overline{c} + \overline{e}) \bullet (c + e)$$
$$\bullet (\overline{a} + \overline{b} + d) \bullet (a + \overline{d}) \bullet (b + \overline{d})$$

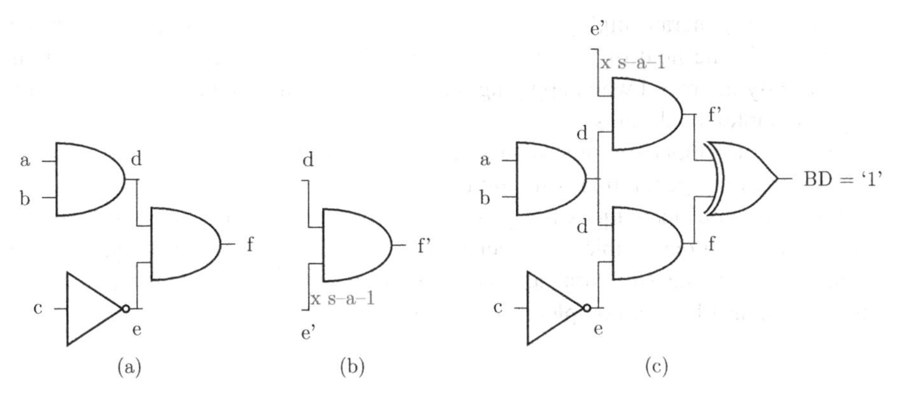

Fig. 3.4 SAT-based ATPG model components for exemplary circuit. (**a**) Fault-free circuit. (**b**) Faulty circuit. (**c**) Boolean difference

The wire e is assumed as the fault site holding a s-a-1 fault. Thus, the faulty value "1" has to be considered for the corresponding signal e' for the faulty circuit F. This faulty value is added to the CNF as a unit clause (e').
Figure 3.4b presents the faulty circuit with one gate only, which is transformed to a CNF representation as follows:

$$\Phi_F = (\overline{d} + \overline{e'} + f') \bullet (d + \overline{f'}) \bullet (e' + \overline{f'})$$

The Boolean Difference circuit BD is realized by appending a Miter and, hence, the difference is enforced by adding the unit clause (BD). This Miter holds for the considered exemplary circuit exactly one XOR gate since the original circuit had just one PO.

$$\Phi_{BD} = (\overline{f'} + \overline{f} + \overline{BD}) \bullet (f' + f + \overline{BD}) \bullet (f' + \overline{f} + BD) \bullet (\overline{f'} + f + BD)$$

The overall instance is built as follows:

$$\Phi = \Phi_G \bullet \Phi_F \bullet \Phi_{BD} \bullet (e') \bullet (BD)$$

The following assignments satisfy Φ:

$$a = 1,\ b = 1,\ c = 1,\ d = 1,\ e = 0,\ f = 0,\ e' = 1,\ f' = 1,\ BD = 1$$

The test T for detecting a s-a-1 at wire e is extracted from the satisfying solution:

$$T = \{a = 1,\ b = 1,\ c = 1\}$$

The SAT-based ATPG can also be combined with an optimization SAT-based approach. This combination allows incorporating new optimization objectives to, for instance, generate highly compact test sets [Egg+16b] or to address arising challenges in the field of power-safe testing [EMW16]. However, the run-time is significantly increased when applying these optimization SAT-based techniques on highly complex IC designs.

In a direct comparison of structural as well as SAT-based ATPG techniques, the latter one allows generating test stimuli for even hard-to-detect faults. In contrast to this, structural techniques are generally not capable to process these hard-to-detect faults in a reasonable time. Generally said, the SAT-based technique is more robust. In turn, the structural techniques require less run-time when processing circuit designs of lower complexity with large fault lists.

3.5 Bounded Model Checking

BMC is a technique, which has been initially designed for functional verification of digital circuits, i.e., proving or disproving temporal properties [DAC99]. A bounded number of time frames of a sequential circuit is checked against a specification mostly in terms of a Linear Temporal Logic-property [Bie+99b, Bie+99a]. However, practically BMC is used for identifying functional failures rather than proving the absence of those.

In [Bie+99b], the BMC problem has been translated into a SAT problem. Thus, efficient SAT solvers are used to solve a BMC instance leading to a practically efficient technique. The determined solutions of the SAT-based BMC instance [Bie+99a] are then used to identify the specific states, which describes the actual *translation* of the Linear Temporal Logic properties into a SAT instance. The orchestration of effective SAT solvers allows verifying industrial-sized microprocessors like POWER5 [Vic+05]. To solve the generated BMC instance by a SAT solver [ES04], the instance has to be unrolled starting from $l = 0$ up to the designer-defined threshold, as stated in Theorem 3.12. The translation of a BMC problem into a SAT instance is done by unrolling the circuit up to a specific time frame applying the technique proposed in [KL93]. The resulting formula can be translated into a CNF in polynomial time with respect to the number of variables and operators by using the encoding technique as described in the previous section.

Definition 3.12 Given a FSM $M = (I, T, S)$ with

- I being a predicate of the initial state,
- T being the transition relation,
- S being the state space, and
- p as given temporal property,

the BMC problem is formulated as follows:

$$\text{BMC} = I(s_0) \bullet \bigwedge_{0 \leq i \leq l} T(s_i, s_{i+1}) \bullet P(s_l)$$

The variable l defines the number of time frames to be unrolled. To cover all reachable states, l needs to be set to the sequential depth of the FSM—often not feasible practically. Hence, l is usually chosen with a proper value by the designer. Therefore, a trade-off between run-time and accuracy is employed.

BMC determines a path of states $s_0 \rightarrow s_1 \rightarrow \ldots \rightarrow s_l$ from the initial state $s_0 \in I$ and $(s_i, s_{i+1}) \in T$ with $0 \leq i < l$ such that the terminal state s_l violates or satisfies a certain property. Once the BMC instance becomes satisfiable, the property is disproved and a counterexample can be extracted from the SAT instance. More precisely, the counterexample is extracted from the determined path, which provides valuable information for debugging. If the BMC instance is not satisfied, the property holds with respect to l. Adjusting the upper bound for l is a crucial task and an open research problem. If the bound is set too low, the BMC model will potentially not reveal the violation of the formulated properties. However, if the bound is set to a value, which is (unnecessarily) high, the run-time will explode since the BMC with its Linear Temporal Logic properties is a computationally hard task that is PSPACE-complete as shown in [SC85].

3.6 Finite State Machine

A FSM is a mathematical model of computation representing a system with a finite number of states. Such a FSM is in exactly one state at a time and the system's response to given inputs are modeled as transitions. These transitions describe valid changes in the system's state as given in Theorem 3.13.

Definition 3.13 A FSM is defined by a tuple $M = (I, S, T)$, where I describes the set of initial states $I \subseteq S$, S represents the state space of the circuit, and T defines the transition relation. A transition relation $T(s, s')$ evaluates to true if there is at least one transition from state s to state s' [Cla+18]. The set of reachable states $S^* \subseteq S$ contains those states that are reachable from an initial state in an arbitrary number of steps.

The concept of FSM is widely used including the field of digital circuit design since the functional behavior of a given sequential circuit can be efficiently represented by a FSM. More precisely, this representation is often used in fields like the formal verification since tasks like the reachability analysis have to be performed, which is relatively easy to conduct a FSM. A corresponding FSM can be generated out of the sequential circuit—as introduced in Theorem 2.2 on page 12—by processing the fan-in as well as fan-out cones of the FFs since the FFs determine the circuit's state. Consequently, a FSM models the circuit's behavior over a number of time frames with respect to the inputs by incorporating the transition relation, i.e., the state-to-state transitions are conducted. This procedure is invoked to determine information about the reachability of the circuit to conclude the correctness of the circuit's functional behavior and, hence, to exclude any erroneous behavior. The set of reachable states $S^* \subseteq S$ contains those states that are reachable from an initial state in an arbitrary number of steps.

3.7 Binary Decision Diagram

A BDD [Ake78] is a graph-based model, which represents, for instance, a Boolean function or a circuit [Lee59] and consists of nodes and edges. The traversal of a BDD is performed in a directed way starting from the root, a node without any predecessors, to a leaf, which is represented by a terminal node, i.e., a node without any successors but holding the value "0" or "1." Hereby, every node—except the leaves—has exactly two outgoing edges, which represents the binary values "0" and "1." Different techniques for minimization exist, which allow reducing the number of nodes and, hence, the size of the BDD. A *reduced* BDD is determined by applying minimization techniques like [DDG00, Sch+99].

One important criterion of the efficacy of a BDD concerns the assumed order of variables when generating the BDD out of, for instance, a Boolean function. If the variables are strictly level-wise arranged, the BDD is said to be ordered. In combination, a reduced and ordered BDD—a ROBDD—is generated, which is a canonical representation of a Boolean function with respect to the assumed variable order. For this reason, BDD is suitable to manipulate (and optimize) the Boolean function [Bry86, DDG00, Sch+99], as applied in Chap. 8 of this book. Thus, a BDD can, on the one hand, be generated out of a truth table and, on the other hand, be synthesized into hardware by applying the BDD-to-MUX synthesis technique of [Som16]. Besides this, a BDD allows evaluating a given function by merely traversing the structure in a directed way. The Boolean function evaluates to the value of the terminal node that is reached at the end of the BDD traversal.

Part II
New Techniques for Test, Debug and Reliability

Chapter 4
Embedded Compression Architecture for Test Access Ports

Since recent years, the profile of functional requirements of ICs has been significantly changing. The ICs are no longer meant to be applied for dedicated functions only but they are designed to address several comprehensive tasks at once. This leads to complex SoC designs including several nested modules and, thus, to a large transistor count. The increased complexity results in a high risk of physical defects accrued while manufacturing. Not only for this reason production tests are indispensable. Besides these production tests, the high modularity of the SoC designs strictly requires extensive board and in-field testing capabilities, which inevitably lead to higher test complexity and to the problem of limited test access to the nested modules. This is since the overall number of pins at the chip-level is limited and, consequently, only a subset of IO pins from the modules can be routed to the chip-level.

Ensuring the accessibility of each and every of the nested modules forms a challenging task. Dedicated test access mechanisms are embedded into the design providing specific access interfaces. The IEEE 1149.1 Std. specifies such a TAP, which is commonly used within industrial designs. This standardized TAP provides capabilities to transfer test data to the CuT. However, one major shortcoming of IEEE 1149.1 concerns the serial interface for the data transfer, which leads to a high TAT, since the IEEE 1149.1 typically operates at a relatively low test clock frequency. In addition, the high TDV and debug data volume require the use of a considerable amount of tester memory, which is often not feasible in a system environment in the field or for rapid post-silicon debug using IEEE 1149.1 and, thus, can lead to complexity margins in the test procedures. This circumstance increases the test costs, which claim a high share of the overall production costs.

To tackle the shortcoming of the state-of-the-art approaches, this book proposes VecTHOR, a new embedded compression technique to be integrated into IEEE 1149.1-compliant TAP controllers. In particular, an architecture has been developed to take advantage of a codeword-based compression technique, which is dynamically configurable and, therefore, applicable even on heterogeneous and

S. Huhn, R. Drechsler, *Design for Testability, Debug and Reliability*,
https://doi.org/10.1007/978-3-030-69209-4_4

high entropic test data. Furthermore, VecTHOR requires only a slight overhead in hardware and no additional pins at the chip-level. VecTHOR uses a code-based compression technique, which can involve fixed codewords as well as dynamically configured codewords for handling heterogeneous test data. VecTHOR can be applied to reduce the TDV of existing test sequences measurably without effecting the logical test behavior itself. Thus, no re-verification of this test data is required and the existing test data have to be retargeted off-chip only once.

Several experiments show that VecTHOR achieves a significant TDV reduction of about more than 25% on high entropic test data and nearly 50% on industrial test data. These experiments show that this technique allows reducing the overall TAT when processing fully specified test data of industrial designs by up to 20%. In reverse, a slight TAT overhead is introduced when processing high entropic test data. Besides this, the embedded compression hardware causes only a negligible overhead in the design itself, which has been proven by synthesizing the extended design.

This chapter is organized as follows: First, the related works are discussed in Sect. 4.1. Section 4.2 presents the compression architecture of VecTHOR including the extension of the FSM, the dynamically configurable embedded dictionary, the codeword-based decompressor as well as an example of VecTHOR being applied. In Sect. 4.3 a retargeting framework is drafted, which processes arbitrary test data effectively by realizing a heuristic retargeting procedure. The experimental setup is described in Sect. 4.4 and the conducted experiments are presented in Sect. 4.5. Finally, a summary and an outlook to future work are given in Sect. 4.6.

4.1 Related Work

Various works have been published in the literature pursuing different approaches that focus on the demands of ATPG, *Embedded Deterministic Test* (EDT) [Raj+04], or of scan chain data compression [JT98, WP02]. Generally, EDT achieves a very high compression ratio, which mostly scales with the share of X-values in the incoming test data. This means that the resulting compression ratio varies strongly depending on the test data and, hence, EDT is not suitable for fully-specified test data as given in the field of functional verification. Furthermore, techniques like EDT require access to designated IO pins at the chip-level. In the case of SoC designs with a large number of nested modules, these IO pins are mostly not accessible at the chip-level. Consequently, these compression techniques cannot be applied during a board or in-field testing using dedicated debug equipment.

Different hardware modules were developed in the past taking advantage of compression methodologies, which are most well-known from the field of software compression. It follows a brief overview of the published methods for TDV compression of ATPG test data. The authors of work [ICM99] proposed a static encoding, which was extended by taking advantage of a static run-length encoding [JT98]. This run-length encoding allows handling repeating bits of the test data efficiently. The proposed method of paper [JGT99] uses the well-known Huffman-Encoding [Ila+13]. A powerful LZ77-algorithm is proposed in [WP02], which is based on a dynamically growing dictionary. A multi-level dictionary has been proposed in [DM13], which requires certain levels of computations.

However, these techniques introduce a significant hardware overhead and lack in a seamless integration into an existing test data flow. The authors in [CC01] introduce a new way of encoding, the Golomb-Code, which takes advantage of previous parts of the test sequence. Additionally, a powerful method of compressing multiple scan chains by a dictionary-based approach was published in [WTH04]. Reference [MK06] proposes a new compression technique, which works independently of X-values and, hence, no X-values have to be injected by the ATPG tool before transmitting the test data.

Concurrent-JTAG [Cla06] was designed to accelerate the functional verification by using TAP structures and is already realized by the industry. However, Concurrent-JTAG requires highly modified devices regarding their TAP controllers so that the compliance with standardized JTAG is no longer ensured. Furthermore, two additional IO pins must be embedded in the top-level. Concurrent-JTAG achieves the speed-up by parallelization necessitating structural requirements for the CuT as well as for test sequences. IJTAG [14] is a technique to establish large on-chip test networks, including the required retargeting of the test data. This technique has been recently addressed by various research works since it holds great potential for the arising challenges in the field of the next generation IC design. Hereby, IJTAG does not contain the chip-level interface but is typically meant for using JTAG TAP controllers. Depending on the size and the structure of the resulting IJTAG network, a certain amount of additional configuration data (for the test network) has to be transferred to the network prior to the actual test data.

Additionally, the authors in [Jia+12] propose a compression scheme for the configuration bitstreams of field programmable gate arrays, which have to be integrated in-between the TAP and the core. Since such a bitstream file is dominated by tailing zeros characteristically, this hardware is based on a Run-Length Encoding, which is suitable for this specialized application field only.

This book focuses on a serial test data transfer into the CuT via a centralized JTAG controller: An extended TAP controller should be developed, which applies a newly proposed compression technique on incoming data. This new technique provides the full legacy support for standardized JTAG operations without increasing the complexity of the controller significantly. The proposed compression architecture can be applied to any data, e.g., fully-specified functional verification test data, for which the majority of previously published techniques are not applicable.

4.2 Compression Architecture

This section describes the essential components of VecTHOR, which are meant to be integrated into an arbitrary IC design that already holds an IEEE 1149.1 TAP controller. Their integration allows to take advantage of a significant TDV reduction, both enabled by VecTHOR.

The general idea of VecTHOR is summarized as follows:
(Prior to the test application, the following steps have to be conducted on the basis of the a-priori known test data once.)

- Analyzing the test data once and identifying the most frequently occurring bit sequences.
- Generating a configuration of an embedded dictionary, where single codewords are configured with most frequently occurring bit sequences.
- Retargeting the incoming test data by replacing all occurrences of the bit sequence with the newly introduced codewords, which generate compressed test data.

(The remaining steps have to be conducted per IC test.)

- Transferring the determined configuration to the chip and applying it to the embedded dictionary introduced by VecTHOR.
- Transferring the compressed test data to the chip and decompressing (restore) the original by the codeword-based decompressor without any loss.

The essential components of VecTHOR are briefly described in the following.

1. The standardized IEEE 1149.1 TAP controller has to be extended such that it is possible to activate the newly embedded compression technique and, otherwise, to isolate the modified datapath completely. This isolation ensures that the full legacy compatibility is given.
2. An embedded dictionary has to be designed, which forms the core of the codeword-based compression, and a mapping function has to be implemented, which projects the codewords to datawords and vice versa.

In addition to these essential components, two significant extensions are designed and seamlessly integrated into VecTHOR to further enhance the effectiveness as follows:

3. A mechanism is implemented, which allows configuring the embedded dictionary dynamically such that the initially configured codewords (by synthesis) are modified. By this, it is possible even to process heterogeneous test data fractions effectively.
4. Certain run-length encoding capabilities are introduced to VecTHOR, which allows to repeat the previously received codeword for an arbitrary length and, thus, addresses homogeneous parts of the test data effectively.

One critical design criterion, which has to be necessarily considered, concerns the resulting hardware overhead of VecTHOR. Since a standardized IEEE 1149.1 introduces only manageable hardware overhead, the implementation of VecTHOR has to be lightweight such that only the same order of additional hardware resources are introduced.

A retargeting procedure is invoked to pre-process existing test data off-chip once. This procedure determines a beneficial *Configuration* ($\mathcal{C}$) for the DDU and generates the *Compressed Test Data* ($\mathcal{D}$) with respect to the recently determined configuration of the embedded dictionary. On the one hand, existing *Uncompressed Test Data* ($\mathcal{I}$) can be easily processed by the developed retargeting framework without any need of regeneration or revalidation and, on the other hand, an existing industrial flow[1] can be directly extended by this retargeting procedure.

The proposed embedded compression flow is shown in Fig. 4.1 and can be separated into three different steps as follows:

1. A retargeting framework is applied on the uncompressed test data $\mathcal{I}$, i.e., the original incoming test data, to determine the configuration of the embedded dictionary and to generate the corresponding compressed test data $\mathcal{D}$.
2. The embedded dictionary is dynamically configured during a preloading phase with the recently determined configuration $\mathcal{C}$.
3. The last step concerns the transfer of the compressed test data and its lossless decompression and, finally, the test application.

4.2.1 Extension of TAP Controller

One essential component of JTAG is the TAP controller implementing the standardized interface protocol, which further contains components like an instruction decoder. It is of uttermost importance that all performed changes are entirely transparent towards the legacy compatibility with IEEE 1149.1 since JTAG has been established as the state of the art.

Commercially available IEEE 1149.1 devices often implement an instruction set by using 4 bits leading to 16 available instruction codes (opcodes). Overall, the standard defines at least 4 instructions, which have to be implemented for conformity reasons and, hence, 12 opcodes remain unassigned. The remaining opcodes can be individually implemented, depending on the given requirements.

VecTHOR introduces two additional instructions by occupying the opcodes `0110` and `0100`, which are representing the new JTAG instructions compr_data and compr_preload and are both not defined in IEEE 1149.1 .

[1] Such a seamless integration has been exemplarily realized by implementing certain input-filters to process STIL formatted data from the TCL environment directly.

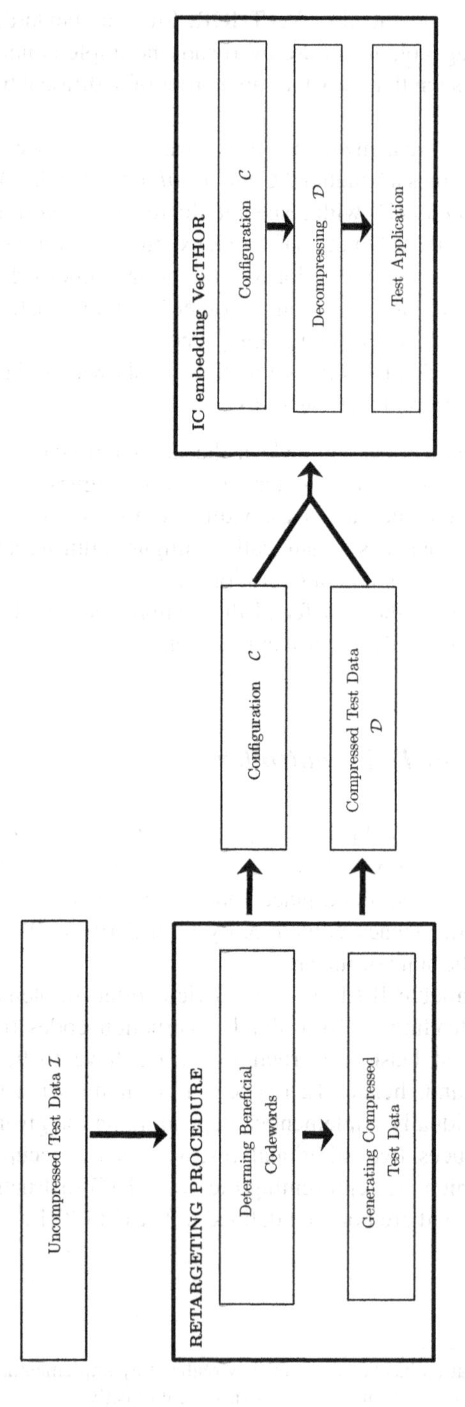

Fig. 4.1 Overall compression flow of VecTHOR

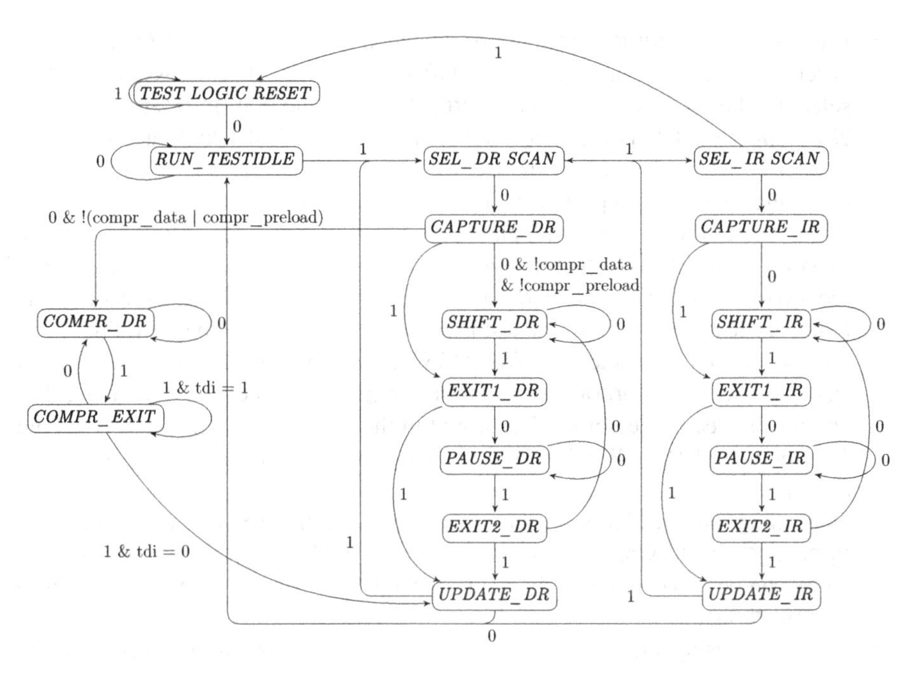

Fig. 4.2 FSM of compression-based IEEE 1149.1: Test Mode Select @ edges

compr_preload allows uploading the determined configuration of the embedded dictionary.

compr_data activates the embedded compression technique or deactivates it if already activated.

If neither compr_data nor compr_preload instruction is loaded, the FSM is traversed following IEEE 1149.1 as shown in Fig. 2.9 on page 25 and the processed data is not affected by any decompression. For the case that instruction compr_data or compr_preload is selected, a TRST trigger—or even unloading this instruction by a further execution—re-enables the normal operation mode. In contrast to the regular FSM as used in JTAG, the modified FSM description contains two additional states compr_dr and compr_exit as presented in Fig. 4.2. Both states are only reachable if the compr_data instruction has been selected.
VecTHOR provides the following two types of compression:

compr utilizes the codeword-based compression technique in combination with the dynamic configurable embedded dictionary.

μ-compr further extends this compr approach by incorporating certain run-length capabilities to address repeating bit sequences effectively. This technique allows decompressing the previously received codeword multiple times without retransfer.

The transfer of compressed data can be described by the following five steps separated by the point in time t_x (measured on rising clock events):

1. If the instruction compr_data has been loaded within the Load_IR phase, TMS is set to "1" to reach the SEL_DR_SCAN state at t_{i-3}. TMS is assigned to "0" selecting the successor's states capture_dr at t_{i-2} and compr_dr at t_{i-1}.
2. *Test Data In* (TDI) is captured on a rising edge of Test Clock at time t_i and is interpreted as a chunk of the overall *Compressed Data Word* (CDW). This capturing process is repeated n times up to t_{i+n} as long as the FSM remains within the compr_dr state. Meanwhile, every received chunk is stored into an exposed Compress Register. The length of the compress register determines the *Maximum Codeword Length* (MCL) l_{CR}. As a sanity check, exceeding this boundary leads to a reset of the test logic, i.e., $l_{CR} \leq n$ is no longer valid.
3. Afterwards, to complete a CDW, TMS has to be changed to "1" causing a state transition to compr_exit. Within compr_exit state, the TAP controller applies the decompressor on data stored in the compress register and writes the *Uncompressed Data Word* (UDW) to the exposed TDR within one test cycle t_{i+n+1}.
4. Depending on the TMS signal at time t_{i+n+2}, the FSM reaches the state update_dr by leaving compr_dr phase with $TMS = 1$, otherwise a further CDW can be processed directly by driving $TMS = 0$, so the FSM passes over to compr_dr again.
5. In case of using the μ-compr technique while compr_data is selected, the transition between compr_exit and update_dr with $TMS = 1$ at t_{i+n+2} has to be enriched by evaluating the TDI signal: If $TDI = 0$ holds, then nothing changes. If $TDI = 1$ holds, then the state compr_exit will be passed through a loop to itself over and over again as long as $TDI = 1 \wedge TMS = 1$. Every cycle iteration decompresses the last received CDW again and writes the mapped UDW to the exposed TDR.

Figure 4.3 demonstrates an exemplary transfer of the compressed dataword "101," which is restored by the mapping function to "10111001." In fact, the implementation of μ-compr triggers the two TDI-controlled state transitions by signal changes on TDI instead of signal values as above drafted. This allows reducing the additional overhead regarding the control data when invoking μ-compr.

4.2.2 Codeword-Based Decompressor

Another important aspect of VecTHOR concerns the decompressor, which consists of a mapping function and of an embedded dictionary. The mapping function projects each and every of the CDW to an UDW uniquely with respect to the configuration of the embedded dictionary as stated in Theorem 4.1.

Definition 4.1 The mapping function Ψ is defined as $\Psi(CDW^c) \rightarrow UDW^u$ realizing a projection of one successfully received CDW of length c with $c \leq$

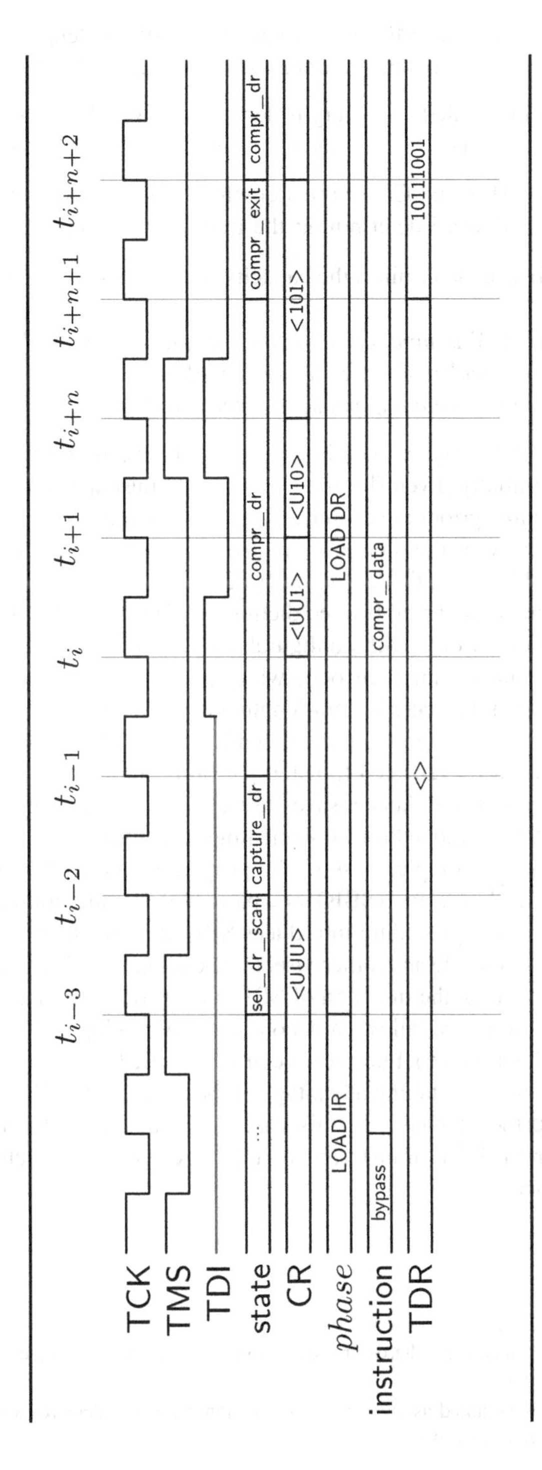

Fig. 4.3 Timing diagram of VecTHOR

Maximal-Codeword-Length (MCL) to a specific UDW of length[2] u with $u \in \{1,\ 4,\ 8\}$. Here, 8 is said to be the *Maximum Dataword Length* (MDL).

Definition 4.2 An embedded dictionary is defined as a $N \times 2$ matrix[3] consisting of N pairs of a codeword (first column) and a dataword (second column).

Definition 4.3 The Maximal-Codeword-Length (MCL) equals the maximal bit length of a single codeword determining the number N of unique codewords by $N = \sum_{i=0}^{MCL} 2^i$ and, hence, determines the capacity of the embedded dictionary.

Lemma 4.4 *Each CDW is uniquely contained in the dictionary, binary encoded using* $1 \ldots MCL$ *bits and initially associated with a certain UDW by synthesis setting or set within the preloading phase, i.e., by a configuration.*

The embedded dictionary holds all associations between codewords and datawords, which are initially given (by the synthesis setting) and, more importantly, dynamically configured prior or even during the test data transfer (see Theorem 4.2). The overall capacity N of the dictionary is determined by the MCL parameter as stated in Theorem 4.3. Thus, the larger the MCL, the more hardware resources are required for implementing the embedded dictionary. The embedded dictionary consists of N pairs of codewords with configurable datawords, where each codeword occurs exactly once and is binary encoded with $1 \ldots MCL$ bits (see Theorem 4.4).

VecTHOR incorporates certain run-length encoding capabilities namely μ-compr, which allows decompressing the recently received CDW on-chip multiple times without the need for repeated transfer. μ-compr allows to address repeating bit sequences effectively and, hence, saves test cycles as well as test data volume. The empty CDW "$\varnothing$" is utilized for implementing μ-compr.

It is required for the completeness of the proposed codeword-based technique that these *Single Bit Injections* (SBIs) are part of the configuration (see Theorem 4.5). The retargeting procedure introduces SBIs at a specific position when no other replacement is possible and, otherwise, this position would remain uncovered. Without loss of generality, the insertion of SBIs allows to process any arbitrary test data successfully and, hence, allows a decompression on-chip without any loss of information (see Theorem 4.6). Furthermore, it is possible to dynamically configure the embedded dictionary by taking advantage of the developed DDU in conjunction with the compr_preload phase. In this case, a pre-determined configuration C as stated in Theorem 4.7 is transferred to DDU performing the changes to the embedded dictionary.

[2]These lengths are assumed due to hardware restriction; however, the technique is applicable for further lengths in principle.

[3]A line of the matrix is assumed as an entry of the dictionary. The codewords are interpreted as CDWs and the datawords as UDWs.

Lemma 4.5 *UDWs with a length of 1 are called SBIs, which are introduced to ensure the completeness of the compression and, hence, to allow a lossless compression in conjunction with a later decompression of arbitrary bit sequences.*

Lemma 4.6 *Given the mapping function Ψ and configuration $\mathcal{C}$ that holds two CDWs for the SBIs cdw_i and cdw_j, such that $\Psi(cdw_i) = 0$ and $\Psi(cdw_j) = 1$. Then, every possible sequence of incoming data $\mathcal{I}$ can be represented by a sequence of CDWs, i.e., the compressed test data $\mathcal{D}$. At least, this can be achieved by simply using the cdw_i or cdw_j successively.*

Definition 4.7 A configuration $\mathcal{C}$ consists of N entries (by following Theorem 4.2), which is applied on an embedded dictionary to reconfigure the mapping of specific codewords (CDWs) to certain datawords (UDWs).

Lemma 4.8 *A compressed test data $\mathcal{D}$ and a configuration $\mathcal{C}$ are given. Then, the original test data $\mathcal{I}$ are restored if the mapping function Ψ (using $\mathcal{C}$) is applied on $\mathcal{D}$, i.e., every CDW contained in $\mathcal{D}$ is sequentially decompressed.*

Example 4.9 An exemplary dictionary with MCL of 3, which is a good trade-off, leads to the following $\sum_{i=0}^{3} 2^i = 15$ binary encodings:

- "∅," "0," "1," "00," "01," "10," "11"
- "000," "001," "010," "011," "100," "101," "110," "111"

These entries are configured by synthesis with pre-initialized UDWs except the empty CDW "∅" that is used for the μ-compr as shown in Table 4.1.
(Note that the index column No. is given for the sake of readability only. For demonstration purposes, the Benefit β value is also shown in this example.

Table 4.1 Exemplary weighted mapping function Ψ, $MCL = 3$

No.	CDW	UDW	Benefit β
0	∅	CDW_{t-1}	–
1	0	1	0
2	1	00000000	7
3	00	1111	2
4	01	0101	2
5	10	0110	2
6	11	0	−1
7	000	01010101	5
8	001	1010	1
9	010	0000	1
10	011	10101010	5
11	100	1000	1
12	101	1001	1
13	110	0001	1
14	111	11111111	5

However, this is neither a part of a configuration nor contained in the embedded dictionary.)

An exemplary dictionary is shown in Theorem 4.9, which contains two CDWs with exposed functions—"0" as well as "11"—representing the SBIs. As soon as the configuration is executed, the compressed test data $\mathcal{D}$ are CDW-wise transferred to the chip and decompressed to restore the original test data $\mathcal{I}$ successively (see Theorem 4.8).

Achieving a high compression ratio is essential to gain a significant TDV. Thus, a mechanism is required, which allows measuring the savings (in the sense of bit count) when introducing a certain replacement. These savings are interpreted as the Benefit and represented as a scalar value β, which is determined by a cost function $B(CDW^c) \rightarrow \beta$ for every dictionary entry individually (see Theorem 4.10).

Definition 4.10 The cost function B is defined as $B(CDW^c) \rightarrow \beta$ determining the benefit of replacing udw_i of length u by a cdw_i of length c with $\Psi(cdw_i) = udw_i$ as $|u - c|$.

Figure 4.4 shows a partial block diagram of the modified TAP controller, which includes the main components like the decompressor, a TDR as well as a functional logic block representing an exemplary data sink, to demonstrate the seamless integration and the full compliance with IEEE 1149.1. Two multiplexers[4] are introduced for decompressing the CDW, which is stored in the exposed compr_reg register and drives the select line of the multiplexer. Depending on the length u of the UDW, the resulting data flow differs. The output of UDWs with $u = 1$ (meaning a SBI) is connected with the serial memory interface, as shown in the upper half of the diagram. The bus output of UDWs with $u = 4$ or $u = 8$ are both wired to the parallel memory interface. Both the serial as well as the parallel bus data are written to introduce FFs as soon as the FSM reaches state compr_exit as described above. These FFs synchronize the decompressed test data with respect to the test clock. The serial write operation is performed by the TDR when write enable (denoted as WE) is set and, alternatively, the parallel operation is performed when both, the WE as well as the parallel data in write enable (denoted as pdi_we) are set. The interconnections between the TDR and functional logic block (denoted as FL) are sketched to demonstrate an exemplary use-case. The diagram includes further components of the regular JTAG concerning the bypass instruction in conjunction with the regular shift_dr state.

[4]Note that the actual implementation applies certain merge operation to avoid redundant multiplexer structures.

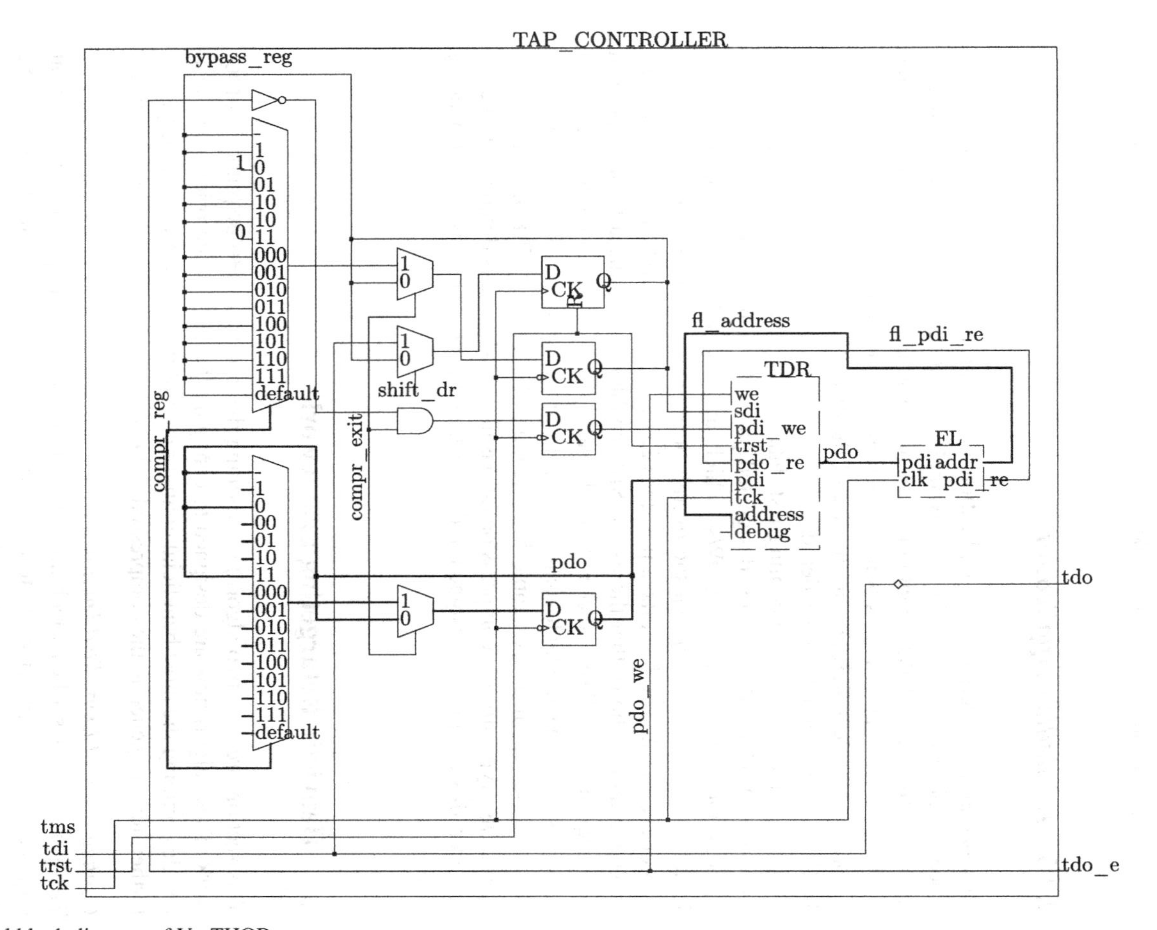

Fig. 4.4 Partial block diagram of VecTHOR

Table 4.2 Applying compr technique on test data

Byte index	0	1	2	
Bit index	0 1 2 3 4 5 6 7	0 1 2 3 4 5 6 7	0	1
Uncompressed test data	$\underbrace{0\,1\,0\,1}\,\underbrace{1\,0\,1\,0}$	$\underbrace{0\,1\,1\,0}\,\underbrace{1\,1\,1\,1}$	$\underbrace{0}$	$\underbrace{1}$
Compressed test data	01 001	10 00	11	0

4.2.3 Exemplary Application

This paragraph describes the application of the proposed compr technique to process arbitrary test data. For this example, the compr technique is applied to an uncompressed test data containing 18 bits. The exemplary mapping function Ψ of Table 4.1 is assumed as well as the drafted retargeting scheme presented by Algorithm 3 in Sect. 4.3. Due to the size of the example, neither the μ-compr nor the DDU configuration is considered.

Every single bit of the incoming test data is assigned to a byte and bit index. As shown, VecTHOR identifies designated parts of the uncompressed vector and encodes them with suitable CDWs, which are highlighted by underbraces in Table 4.2. Due to the fact that the overall bit number is not completely divisible by 8 or even 4, the two last bits at positions 2.0 and 2.1 have to be handled as SBIs.

This example shows that the use of SBIs enables processing arbitrary test data to a sequence of CDWs successfully. Finally, this compr technique generates compressed test data, which contains only 12 bits. A compression ratio of nearly 34% is achieved in this example by extending the existing TAP mechanism by VecTHOR. Typically, the compression ratio is higher when considered real test data; however, this example is designed to demonstrate the completeness by introducing SBIs.

4.3 Heuristic Retargeting Framework

This section presents two algorithms forming the developed Retargeting Framework. These algorithms are designed to configure the DDU once prior to the test, i.e., determining the most beneficial configuration of the embedded dictionary, and, furthermore, to generate the compressed test data by processing arbitrary test data.

Configuration of Embedded Dictionary

This paragraph describes a technique to identify the most frequently occurring bit sequences of an a-priori given test data set as stated in Algorithm 2. At first, the original vector Ω is serialized and processed such that all sub-sequences with a length of u, i.e., a supported UDW length, are handled separately (line 2 ff.). The occurrences of possible UDW permutations are counted within a data container Γ. Afterward, the entries are post-processed by removing existing intersections (lines 5

Algorithm 2 Retargeting algorithm: configure

Require: Ω, Ψ
1: $\Gamma := ()$ {Empty data container for counting occurrences}
2: **for all** (i, j) s.t. $\omega_i \dots \omega_j \subseteq \Omega \wedge i \leq j \wedge \text{distance}(i, j) \in u$ **do**
3: increaseOccurrencesCtr((ω_i, ω_j)) in Γ
4: **end for**
5: **for all** $\gamma \in \Gamma$ **do**
6: $(\omega_i, \omega_j) := \gamma$
7: **if** $\exists(\omega_k, \omega_m) \in \Gamma \backslash \{\gamma\}$ s.t. $i \leq k \leq j \vee i \leq m \leq j$ **then**
8: $\widehat{\gamma} := (\omega_k, \omega_m)$
9: **if** calculateGlobalBenefit(γ) $\leq$ calculateGlobalBenefit($\widehat{\gamma}$) **then**
10: decreaseOccurrencesCtr(γ) in Γ
11: **else**
12: decreaseOccurrencesCtr($\widehat{\gamma}$) in Γ
13: **end if**
14: **end if**
15: **end for**
16: sortByDecreasingOccurrencesCtr(Γ)
17: **for all** $\gamma_i \in \Gamma$ s.t. $i \in [0, 2^{MCL})$ **do**
18: emplaceInDDU(γ_i)
19: **end for**

to 15). Within this step, the benefit of all pairwise intersections is evaluated (line 9) and the less valuable one is removed by decreasing the corresponding counter in Γ. Finally, Γ is sorted by a decreasing number of occurrences (line 16) and the most valuable 2^{MCL} ones are emplaced in the DDU configuration (line 17 ff.) by invoking the compr_preload instruction.

Heuristic Retargeting Procedure

This paragraph drafts the retargeting procedure given in Algorithm 3, which applies the most beneficial replacements to the data set resulting in the compressed test data. Besides this DDU configuration set, a further algorithm is required to identify valuable segments of the original test data with respect to the current DDU configuration, which executes their replacements and emits VecTHOR's extended JTAG protocol. The retargeting procedure processes the original vector Ω to a suitable vector $\overline{\Omega}$ as shown in Algorithm 3. Ω consists of N data bits ω_1, ..., ω_N and $\overline{\Omega}$ consists of compressed data bits $\overline{\omega_1}$, ..., $\overline{\omega_C}$ to be stored in the test equipment. The replacements Δ must be calculated to determine $\overline{\Omega}$: Every single replacement $\delta \in \Delta$ maps to an underlying UDW as well as the range $(s_i : s_j)$ within the bit vector Ω, i.e., $\omega_i \dots \omega_j$ getting replaced by δ. This set of replacements can be directly processed into the desired $\overline{\Omega}$, which is transmitted to the TAP controller.

Initially, Δ is calculated in Algorithm 3 as follows:

1. The set $\widehat{\Delta}$ of all possible replacements is determined (line 1).
2. Valuable start points $InitSet$ with a benefit β greater than zero are extracted and sorted by β decreasingly (line 2 f.).

Algorithm 3 Retargeting algorithm: compress

Require: Ω
1: $\widehat{\Delta} := \{ (CDW, s_i, s_j) \mid \Psi(CDW) = \omega_i ... \omega_j \subseteq \Omega \wedge i \leq j \}$
2: $InitSet := \{ \delta \in \widehat{\Delta} \mid B($ getCDW$(\delta)) > 0 \}$
3: $InitSet :=$ sortByDecreasingBenefit($InitSet$)
4: $\Delta := \varnothing$
5: {Processing start points}
6: **for all** $init \in InitSet$ **do**
7: $(CDW, s_i, s_j) := init$
8: **if** $\exists i' \in [i, j] \ : \ s_{i'}$ isCoveredBy(Δ) **then**
9: **continue** {Skip conflicting replacement}
10: **end if**
11: $\Delta := \Delta \cup init$ {Add replacement}
12: **end for**
13: $\overline{\Omega} := \varnothing$
14: {Filling existing uncovered areas in bit vec}
15: **for all** $i' \in [1, N] \ : \ s_{i'}$ is**Not**CoveredBy(Δ) **do**
16: $\delta_{i'} := (\Psi^{-1}(\omega_{i'}), \ s_{i'}, \ s_{i'})$
17: $\overline{\Omega} := \overline{\Omega} \cup \{\delta_{i'}\}$
18: **end for**
Ensure: Every data bit is covered.
19: {Generating sequence of compressed data bit}
20: $\Delta :=$ sortByIncreasingPos(Δ)
Ensure: Next step only in case of μ-compr.
21: $\Delta :=$ mergeConsecutiveIndenticalCDWs(Δ)
22: **for all** $\delta_i \in \Delta$ **do**
23: $\overline{\Omega} := \overline{\Omega}$ **concat** $\Psi^{-1}(getCDW(\delta_i))$
24: **end for**
25: **return** $\overline{\Omega}$

3. All replacements are inserted into Δ as long as they are not conflicting with previously introduced replacements (lines 6 to 12).
4. If the complete coverage of all data bits $\omega_{i'}$ with $i' \in [1, N]$ is not achieved yet, the remaining uncovered bits must be filled. For simplifying the shown algorithm, only SBIs are considered and, hence, the uncovered single bit $\omega_{i'}$ is replaced by a SBI $\delta_{i'}$ with $\beta \leq 0$ (line 16 f.).
5. If the coverage is completed, the set of replacement is sorted and converted into the final vector containing compressed data bits $\overline{\Omega}$ (lines 22 to 24), which finalizes the retargeting process.
6. In case of using μ-compr, the merging step has to be executed just before starting the concatenation (line 21) by substituting specific not-empty replacements with $\varnothing$.

4.4 Experimental Setup

This section describes the comprehensive experimental setup, which has been implemented to evaluate and to validate the developed embedded compression architecture.

The proposed extended TAP controller VecTHOR is fully compliant with the IEEE 1149.1 standard [Moh09], which is an important attribute since JTAG serves as the state-of-the-art and, thus, is widely used. VecTHOR has been solely implemented in Verilog and tested by using a simulation-based technique and by implementing a hardware prototype based on a field programmable gate array device. The latter ensures that the developed hardware modules are synthesizable.

A TDR has been appended to this design for creating a realistic use-case scenario, which provides a serial as well as one byte parallel dual-port memory interface connected to the TAP controller. This TDR acts as the data sink when the new compr instruction is loaded and, hence, the received data are stored in this TDR after being decompressed by the DDU. Additionally, a functional logic block was integrated into this design, which represents an arbitrary CuT.

Two Verilog test benches are utilized for simulating the test data transfer to a CuT for validation of the obtained results: $\mathrm{TB_{LEG}}$ serves as a reference TAP controller and $\mathrm{TB_{COMPR}}$ embeds a TAP controller implementing VecTHOR, both fully compliant with IEEE 1149.1. The $\mathrm{TB_{LEG}}$ is applied for each test case on the incoming test data $\mathcal{I}$ to determine the content of the TDR after the transmission has been completed, i.e., the golden test results. The configuration for the embedded dictionary $\mathcal{C}$ and the compressed test data $\mathcal{D}_i$ are generated by applying the retargeting technique on $\mathcal{I}$. Subsequently, $\mathcal{C}$ is used to configure the DDU and $\mathcal{D}_i$ is transferred by $\mathrm{TB_{COMPR}}$. The data stored in the TDR have to be equivalent with the golden one determined by $\mathrm{TB_{LEG}}$ to pass the validation. Figure 4.5 visualizes the experimental setup.

Different use-cases require heterogeneous test data throughput and, hence, multiple test data sizes starting with 256 bytes (N_1) up to 8192 bytes (N_4) were assumed. All these test data ($RTDR\_N_1$, ..., $RTDR\_N_4$) were generated by a pseudo-random number generator based on the Mersenne Twister algorithm, which determines $\{0, 1\}$ as elements of bit strings. These random bit strings are characterized by a very high entropic such that the lower bound for the compression ratio should be determined [BT07]. For instance, this can be achieved by VecTHOR while processing precompressed test data. Besides experiments on high entropic random test data, further experiments were conducted for fully-specified test data for various industrial circuits containing 35k to 378k nets, which were provided by NXP Semiconductors. The average value of all patterns is used for industrial data.

The retargeting framework has been compiled and all benchmarks have been executed on a Fedora 30 (with gcc 9.2.1) system holding an Intel Xenon E3-1240v2 3.4 GHz processor with 32 GB system memory.

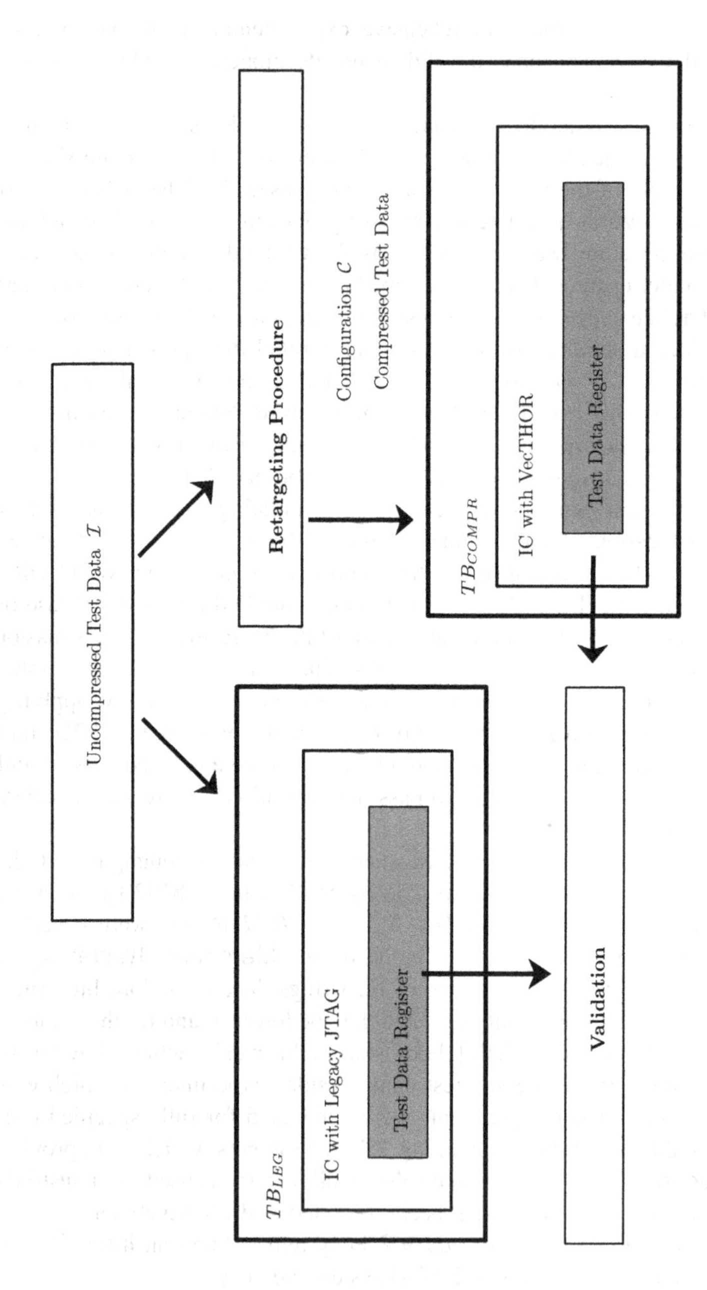

Fig. 4.5 Experimental setup

4.5 Experimental Results

This section evaluates the conducted experiments concerning the high entropic as well as the industrial test data. Three different experiments are conducted for every benchmark as follows: leg refers to the legacy JTAG operation mode, compr means that the new compression instruction is Invoked, and μ-compr takes advantage of the merge-compress technique. Results concerning the reduction of the TDV are given in detail in Table 4.3. The column size [bit] indicates the resulting data volume and column TDV red. [%] gives the data volume reduction ratio. Analogously, Table 4.4 presents the results concerning the number of test cycles (column #data cycles) and the overall TAT reduction (column TAT red. [%]). The additional data volume and test cycles, which are introduced by the applied configuration, are considered in the separated column config. Note that the calculated TDV and TAT reduction consider the configuration of the embedded dictionary.

The experiments clearly demonstrate that the compression and the retargeting is robust when processing different types of data. The run-time of the retargeting processes is very low with less than $10s$ for all conducted experiments. Figure 4.6 presents the TDV reduction in percentage and Fig. 4.7 the TAT reduction in percentage when applying VecTHOR's compr and μ-compr techniques on a different set of test data.

For a visual comparison, the TDV and TAT are plotted as histograms in Figs. 4.6 and 4.7. VecTHOR achieves a TDV reduction between 16.8% to 21.0% for high entropic random data (RTDR_*) by using compr and between 23.3% to 26.3% by using μ-compr technique. The compression ratio is even higher when considering real test data of industrial circuit designs as an input: compr achieves a compression ratio of 29.0% to 45.4% and the μ-compr technique enables an even higher ratio

Table 4.3 TDV reduction of random and industrial circuit designs

	Run-time [s]		Size [bit]				TDV red. [%]	
Test name	∅compr	∅μ-compr	leg	config	∅compr	∅μ-compr	compr	μ-compr
RTDR_256	0.93	1.11	2,048	28	1,675	1,542	16.8	23.3
RTDR_512	1.98	2.16	4,096	28	3,302	2,999	18.7	26.1
RTDR_1024	4.57	4.86	8,192	24	6,450	6,011	21.0	26.3
RTDR_2048	9.25	9.68	16,384	24	13,218	12,085	19.2	26.1
RTDR_4096	19.3	19.9	32,768	48	26,318	24,152	19,8	26.4
RTDR_8192	37.3	38.1	65,536	48	52,795	48,729	19,5	25.7
p100k	2.71	2.95	5,902	44	3,180	2,933	45.4	49.6
p267k	8.45	9.12	17,332	41	9,514	8,814	44.9	48.9
p330k	8.73	9.43	18,012	40	9,895	9,214	44.8	48.6
p35k	1.26	1.42	2,912	38	1,785	1,603	37.4	43.6
p378k	7.54	8.25	15,732	30	10,743	9,783	31.5	37.6
p78k	1.37	1.55	3,148	31	2,205	1,980	29.0	36.1
p81k	1.78	1.99	4,030	35	2,692	2,420	32.3	39.1

Table 4.4 TAT reduction of random and industrial circuit designs

	#data cycles				TAT red. [%]	
Test name	leg	config	∅compr	∅μ-compr	compr	μ-compr
RTDR_256	2,053	31	2,436	2,303	−20.2	−13.7
RTDR_512	4,101	31	4,831	4,528	−18.6	−11.2
RTDR_1024	8,197	27	9,354	8,915	−14.4	−9.1
RTDR_2048	16,389	27	19,119	17,986	−16.9	−9.9
RTDR_4096	32,773	100	38,277	35,851	−17.1	−9.7
RTDR_8192	65,541	100	76,582	72,060	−17.0	−10.1
p100k	5,909	65	4,718	4,471	19.1	23.3
p267k	17,338	62	14,12	13,421	18.2	22.2
p330k	18,016	61	14,659	13,977	18.3	22.1
p35k	2,919	59	2,715	2,533	5.0	11.2
p378k	15,738	51	16,027	15,067	2.1	4.0
p78k	3,154	52	3,326	3,101	−7.1	0.0
p81k	4,035	56	4,033	3,760	−1.3	5.4

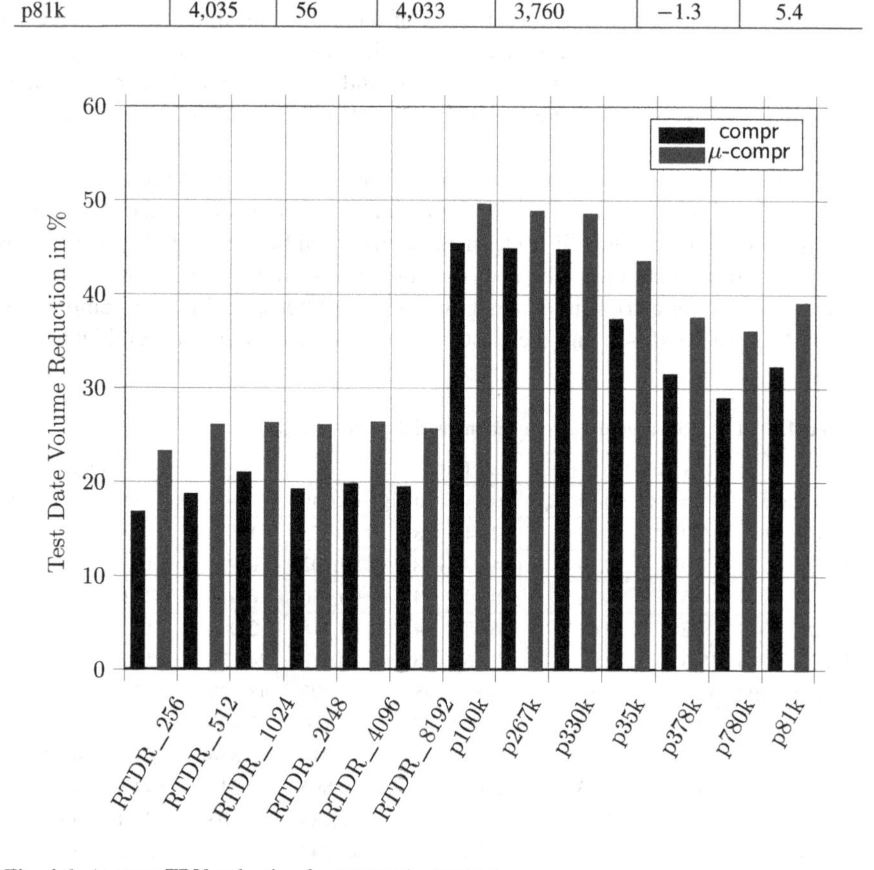

Fig. 4.6 Average TDV reduction for compr (μ-compr)

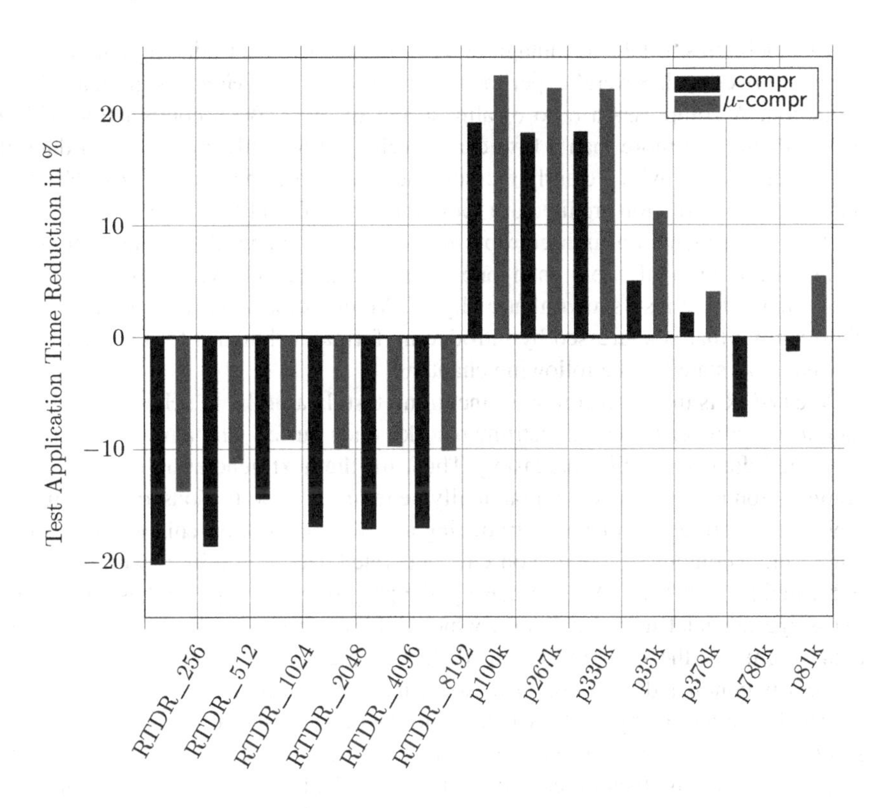

Fig. 4.7 Average TAT reduction for compr (μ-compr)

of 36.1% to 49.6%. Furthermore, a measurable TAT reduction is achieved when processing industrial circuit test data with up to 23.3%. However, further test cycles are introduced by μ-compr when processing high entropic random test data.

The conducted experimental data demonstrate that the μ-compr technique, which incorporates run-length encoding capabilities, affects the retargeting run-time only slightly but achieves a higher compression ratio and lower number of overall needed test cycles than the compr technique. It can be further observed that there is only a low variance in the compression factors meaning that the compression technique is robust over a heterogeneous type of test data.

In summary, VecTHOR allows reducing almost half of the TDV, which directly correlates with the limited memory resources to be allocated at the test equipment.

4.6 Summary and Outlook

This chapter proposed VecTHOR: A new low-cost compression architecture for IEEE 1149.1-compliant TAP controller in conjunction with a suitable retargeting framework. This framework allows processing arbitrary test data effectively. Vec-

THOR facilitates to take advantage of TDV reduction by introducing only a slight hardware overhead. Several experiments have proven VecTHOR's significant TDV reduction: A compression ratio of almost half of the TDV in combination with a TAT reduction of more than a fifth can be achieved for fully-specified test data of industrial designs, which clearly recommends the implementation of VecTHOR as a low-cost compression architecture for compliant IEEE 1149.1 TAP controllers. Further experiments on high entropic test data have shown that an average TDV compression ratio of more than one quarter can be achieved. However, these experiments have also shown that a slight TAT overhead is introduced, forming a shortcoming that is addressed by introducing formal techniques for the retargeting procedure as stated in the following chapter.

VecTHOR is meant to compress incoming test data of the IC. However, with an increasing number of debug instruments like trace buffers, the amount of on-chip generated data is steadily increasing. Thus, for the next generation of IC suitable compression architectures will potentially be required as well. Consequently, future work will continue the idea of introducing a neural network on-chip to apply a fast but accurate compression on the on-chip generated data. Preliminary work has been published in [Huh+18]. More precisely, the proposed technique orchestrates a fast two-stage artificial neural network, which achieves a significant reduction of the data volume of the on-chip generated debug data. Consequently, the retargeting process is conducted by a neural network instead of invoking a heuristic or even a formal retargeting approach, which cannot be performed directly on-chip without synthesizing a dedicated co-processor. The proposed neural network is trained off-chip to keep the lightweight character of the sketched solution since implementing on-chip training capability would require measurably more hardware. Thus, the parameters of the neural network are extracted and directly encoded in the resulting circuit, which is meant to be seamlessly integrated into the existing test/debug infrastructure. First experiments are conducted on a large trace data set, which has been determined by observing randomly selected signals of a state-of-the-art open-source microprocessor implementation, and prove that the proposed scheme works in principle. The significant reduction of the resulting data volume contributes to the availability of more comprehensive debugging mechanisms.

Chapter 5
Optimization SAT-Based Retargeting for Embedded Compression

SoCs are now widely used in the semiconductor industry to fulfill the challenging functional requirements of nowadays application scenarios. This inevitably leads to higher test complexity since comprehensive test scenarios have to be considered for functional verification. Due to the SoC character, further challenges have to be addressed, which concern the accessibility during the test and debug. Typically, test access mechanisms are embedded in the design to ensure the required accessibility of nested modules. The IEEE 1149.1 Std. specifies a TAP, which is commonly used in industrial designs allowing the transfer of test data. Since IEEE 1149.1 provides only a serial interface for the data transfer operating at a relatively low test clock frequency, the bandwidth is strictly limited. Consequently, the transfer of large test data consumes much time and, hence, leads to a high overall TAT, which directly increases the resulting test costs.

VecTHOR has been proposed in Chap. 4 to tackle these shortcomings and to reduce the overall TDV significantly as well as the TAT. More precisely, a complete architecture is developed for taking advantage of a codeword-based compression technique, which is dynamically configurable, and therefore applicable even on high-entropic test data. VecTHOR can be seamlessly integrated into any existing IEEE 1149.1-compliant TAP controller and requires only a slight overhead in hardware and no additional pins at the chip-level. For utilizing this compression technique, existing test data have to be retargeted off-chip only once without any need for regeneration or re-verification.

As clearly shown by the experiments in Sect. 4.5, VecTHOR achieves promising results concerning the TDV reduction as well as the TAT reduction under certain conditions. The effectiveness of compression strongly depends on the selection of the codewords, i.e., the configuration of the embedded dictionary, since the compression technique is mainly codeword-based. In Chap. 4, a heuristic approach has been introduced, which allows processing test data of industrial designs effectively by almost halving the TDV and by reducing the TAT by around 20%. However, the approach is not yet exploiting its full potential, in particular, when

S. Huhn, R. Drechsler, *Design for Testability, Debug and Reliability*,
https://doi.org/10.1007/978-3-030-69209-4_5

processing high-entropic test data. A greedy algorithm is not likely to find optimal codewords since a large search space exists, which is spanned by the binominal equation of the dictionary size choose number of possible codewords. Consequently, the dictionary is not chosen optimal. Thus, the compression effectiveness is reduced and, more importantly, an overhead of test cycles is introduced. This has an adverse impact on the TAT, which leads to an increased transfer time.

The following chapter proposes a new formal optimization-based technique to determine a set of optimal codewords meaning the configuration of the embedded dictionary. The determination of an optimal configuration for a given test data stream is modeled as a PBO problem for which solvers exist that are able to compute appropriate solutions in a reasonable time. In contrast to the greedy algorithm of Chap. 4, the proposed optimization-based technique determines an optimal set of codewords (under the current parameters) and, hence, outperforms the greedy algorithm for hard-to-compress random test data. More precisely, this configuration is used to dynamically configure the DDU, which yields an increased compression ratio and a decreased number of required test cycles. Various experiments are conducted on random test data, on debugging data of a JPEG encoder, as well as on commercially representative functional verification test data for a state-of-the-art softcore microprocessor. Random test data have characteristically a high entropy and, hence, this type of data determines the lower bound on the compression ratio typically [BT07]. The results clearly show that the new optimization-based approach improves the compression results significantly compared to the results of heuristic techniques, as previously proposed in Sect. 4.3. A TDV reduction by up to 37.1% can be achieved for high-entropic data. The TAT overhead of previous techniques has been completely eliminated. In fact, the TAT for functional verification test data can be even reduced by up to 20% compared to the time an uncompressed data transfer consumes.

The structure of this chapter is as follows: Sect. 5.1 describes the DDU while focussing on the importance of a deliberately determined configuration of the embedded dictionary. In Sect. 5.2, the optimization SAT-based model is motivated and the developed modeling scheme to, among others, consider specific hardware constraints are presented in detail. The aggregation of the complete model yielding to the optimization-based retargeting procedure is sketched from an algorithmic point of view in Sect. 5.3. The experimental setup is presented in Sect. 5.4, the conducted experiments are shown in Sect. 5.5, and a summary and an outlook to future work are given in Sect. 5.6.

5.1 Dynamic Decompressing Unit

VecTHOR combines an IEEE 1149.1-compliant test access mechanism with a dynamically configurable codeword-based compression architecture, which also provides run-length encoding capabilities. Instead of transferring the original test data directly to the SoC using the standardized JTAG, the test data are preprocessed by a suitable retargeting framework off-chip once prior to the transfer. This retargeting procedure generates the compressed test data $\mathcal{D}$ consisting of a CDW sequence to be transferred to the extended TAP controller. Afterwards, the CDW sequence is expanded by the DDU on-chip, i.e., the original (uncompressed) test data are restored losslessly.

The DDU holds an embedded dictionary, which stores the mapping between CDWs and UDWs following the definitions on pages 60 and 62. This dictionary is dynamically configurable at any time between two data transfers by invoking compr_preload as described in Sect. 4.2.1 (on page 57). It is required to determine a configuration $\mathcal{C}$ in advance, which is then used to perform a (re-)configuration of the embedded dictionary. The instruction code compr_preload is prepended to the determined $\mathcal{C}$ to start the reconfiguration process, which will then selectively overwrite the images of Ψ.

An exemplary implementation of such a mapping function Ψ was shown in Table 5.1. Following the same encoding as stated in Chap. 4, the CDWs "∅," "0," and "1" hold exposed functions: "∅" encodes the run-length encoding capability, i.e., at the current time t, the last successfully transferred CDW (at time of $t - 1$) is repeated. "0" and "1" are representing SBIs, which ensure that arbitrary test data can be processed.

Tables 5.1 and 5.2 show a simplified example to demonstrate the importance of deliberately selected codewords. Two different configurations $\mathcal{C}^1$ and $\mathcal{C}^2$ are presented in Table 5.1. Here, the mapping between codewords (column CDW) and uncompressed datawords (column UDW) is given. As shown in Table 5.2, $\mathcal{I}$ contains 24 bits and is compressed to a sequence of 6 CDWs by using $\mathcal{C}^1$ and to a sequence of 4 CDWs by using $\mathcal{C}^2$. These CDW sequences contain 15 bits overall ($\mathcal{C}^1$) and 8 bits ($\mathcal{C}^2$), i.e., a TDV reduction of about 37.5% ($\mathcal{C}^1$) and about 66.7% ($\mathcal{C}^2$)

Table 5.1 Example configurations $\mathcal{C}^1$ and $\mathcal{C}^2$ for mapping function Ψ using $MCL = 3$

No.	CDW	UDW @ $\mathcal{C}^1$	UDW @ $\mathcal{C}^2$
1	∅	CDW@$t-1$	CDW@$t-1$
2	0	0	0
3	1	1	1
4	00	1111	01011010
5	01	0101	0110
6	10	0110	0001
7	11	00000000	10010110
8	000	01010101	1100
9	001	1010	1010
10	010	0000	0000
11	011	10101010	10101010
12	100	1000	1000
13	101	1001	1001
14	110	0001	0001
15	111	11111111	11111111

Table 5.2 Application on exemplary test data using configuration $\mathcal{C}^1$ and $\mathcal{C}^2$

Byte index	0	1	2
Bit index	0 1 2 3 4 5 6 7	0 1 2 3 4 5 6 7	0 1 2 3 4 5 6 7
Configuration $\mathcal{C}_1$			
Uncompressed test data	0 1 0 1 1 0 1 0	0 1 1 0 0 0 0 1	1 0 0 1 0 1 1 0
Compressed test data $\mathcal{D}_1$	01 001	10 110	101 10
Configuration $\mathcal{C}_2$			
Uncompressed test data	0 1 0 1 1 0 1 0	0 1 1 0 0 0 0 1	1 0 0 1 0 1 1 0
Compressed test data $\mathcal{D}_2$	00	01 10	11

is achieved. As clearly demonstrated, the selected configuration $\mathcal{C}$ has a substantial impact on the compression ratio. Determining a good configuration is a challenging task due to mutual dependencies and codeword overlaps in the test data stream.

5.2 Optimization SAT-Based Retargeting Model

VecTHOR offers a powerful mechanism for TDV and TAT reduction. However, the challenge of codeword selection for an arbitrary test data stream, in particular, when processing high-entropic test data is yet to be solved satisfactorily. The remainder of this chapter describes the newly proposed technique that is able to tackle this shortcoming. A retargeting procedure is presented, which orchestrates formal optimization-based techniques to determine an optimal configuration for the embedded dictionary and, furthermore, determine the most beneficial sequence of replacements. The proposed technique is adopted on the selected parameters of VecTHOR as presented in Chap. 4, which are assumed to incorporate the hardware constraint. However, the optimization-based model can be easily adopted to arbitrary parameters.

5.2.1 Motivation

The general idea is about formulating the retargeting problem as a formal optimization problem. The proposed retargeting flow is shown in Fig. 5.1, where the original uncompressed test data $\mathcal{I}$ are given as an input. An optimization problem is formulated, which consists of SAT as well as PB constraints and an optimization function. A PBO solver is called to solve the previously generated problem instance. The result is necessarily a satisfying model. This is due to the introduction of SBIs, which ensure the completeness of the retargeting procedure and, hence, any arbitrary test data can be successfully processed. The model parameter are extracted and directly used to determine an optimal configuration $\mathcal{C}$ as well as the targeted compressed test data $\mathcal{D}$. Since the TAP controller is dynamically configurable,

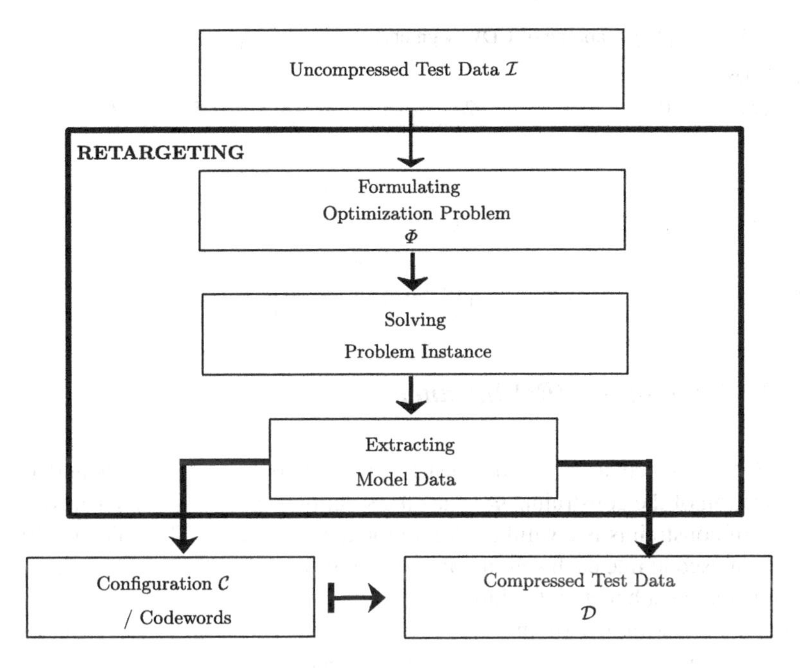

Fig. 5.1 Proposed optimization-based retargeting flow

the calculated configuration $\mathcal{C}$ can be loaded into the controller to decompress the compressed test data $\mathcal{D}$ on-chip, which restores the original uncompressed test data $\mathcal{I}$. This technique allows a TDV reduction of the data, which have to be serially transferred into the TAP controller. This procedure can also be used for partitioned data streams due to the dynamic configuration.
The hardware constraints are considered yielding to the following assumptions:

1. A circuit design that embeds a codeword-based TAP controller, which includes a DDU holding an embedded dictionary consisting of $n = \sum_{i=2}^{MCL} 2^i$ dynamically configurable entries.
2. Three dictionary entries are statically included, i.e.,"$\varnothing$," "0," "1." Again, "$\varnothing$" realizes the run-length encoding capability and both codewords with a length of 1 for representing both SBIs. Consequently, $n-3$ codeword entries are dynamically configurable by the proposed retargeting procedure.
3. Let $u = \{1, 4, 8\}$ be the UDW length supported by the TDR interface. Then, $2^1+2^4+2^8$ possible Boolean permutations exist. Each permutation is a candidate for being included as a codeword in the DDU.
4. The incoming test data sequence $\mathcal{I}$ contains only fully-specified values meaning "0" and "1."

Based on these basic conditions, a formal PBO model has to be built, which allows the automatic computation of valid configurations with an optimal codeword selection. Configuring the embedded dictionary with this optimal configuration yields a significant reduction of the TDV and the TAT.

Table 5.3 Exemplary mapping of UDW segments

Byte index	0							
Bit index	0	1	2	3	4	5	6	7
8C	v_0^{8C}							
4C	v_0^{4C}				v_4^{4C}			
4C		v_1^{4C}						
4C			v_2^{4C}					
4C				v_3^{4C}				
1C	v_0^{1C}	v_1^{1C}	v_2^{1C}	v_3^{1C}	v_4^{1C}	v_5^{1C}	v_6^{1C}	v_7^{1C}

5.2.2 *Generating PBO Instance*

The PBO instance to develop consists of the two components Φ and $\mathcal{O}$. The formulation of the constraints Φ spans the solution space such that each solution to the set of constraints is a valid configuration and vice versa. The objective function $\mathcal{O}$ is then used to rate each solution in terms of its costs. By this, the search is guided towards the most beneficial solution.

First, the meanings of the variables are described, which are related to the problem instance. Since each possible segment in the data stream $\mathcal{I}$ has to be covered by a codeword, a variable v_i^{uC} is used to determine the status of every segment. The parameter i denotes the offset of the segment in $\mathcal{I}$. This offset has to be considered for every UDW length u individually. An example is shown in Table 5.3 listing all possible segments of a 8 bit data stream. The number of possible segments depends on u as follows: Only one segment is possible for $u = 8$, $u = 4$ leads to 5 possible segments and, finally, 8 segments are possible for $u = 1$.

A variable v_i^{uC} is introduced to keep track of the status of a certain segment. More precisely, the variable v_i^{uC} is assigned to "1," if the corresponding segment of length u is meant to be covered by a codeword, said to be active. As a further step, all possible UDW permutations have to be modeled in the problem instance. This is important for the determination of the codeword images, which are currently part of the prospective configuration. Two variables $v_{\mathrm{UDW}}^{\mathrm{2CDW}}$ and $v_{\mathrm{UDW}}^{\mathrm{3CDW}}$ are assigned to every possible UDW permutation to keep track of the active images to be configured. One variable is not sufficient since codewords of different lengths, i.e., length of 2 and 3 bits, have to be distinguished for the hardware requirements. Hereby, the codewords of length 1 are encoded statically since these model SBIs. A total number of $2 \cdot (2^4 + 2^8)$ variables is required. For example, $v_{0001}^{\mathrm{2CDW}} = 1$ means that the UDW "0001" will be replaced by a codeword of length 2. The actual codeword is not relevant for the calculation.

Different constraints have to be generated such that the solving algorithm is able to consistently assign these variables. Overall, the problem instance including these constraints is defined by

$$\Phi = \Phi_{\#\mathrm{CDW}} \bullet \Phi_{\mathrm{uC}} \bullet \Phi_{\mathrm{ME}} \bullet \Phi_{\mathrm{RET}}$$

$\Phi_{\#\text{CDW}}$ guarantees that the maximum number of dictionary entries is not surpassed, Φ_{uC} realizes that all possible segments are covered to ensure the completeness, Φ_{ME} ensures that all bits of the original test data $\mathcal{I}$ are covered exactly once, $\Phi_{\#\text{CDW}}$ guarantees that the maximum number of dictionary entries is not surpassed, and, finally, Φ_{RET} implements the retargeting procedure itself.
The remainder of this section describes the above-listed components of the problem instance sequentially.

Maximum Number of CDWs
Different hardware restrictions exist, which have to be considered when implementing VecTHOR. These restrictions allow only a limited number of dictionary entries and, hence, all solutions have to be excluded from the solution space, which would configure more than the allowed number. In our case, $2^2 = 4$ different entries are allowed for codewords of length 2 and $2^3 = 8$ different entries are allowed for codewords of length 3. This can be modeled as weighted constraints as follows:

$$\Phi_{\#\text{CDW}} = \left(\sum_{i=1}^{272} c \cdot v_i^{2\text{CDW}} \leq 4\right) \bullet \left(\sum_{i=1}^{272} c \cdot v_i^{3\text{CDW}} \leq 8\right)$$

with an equal weight $c = 1$ and x_i the positive literal representing the variable $v_{\text{UDW}}^{2\text{CDW}}$ and $v_{\text{UDW}}^{3\text{CDW}}$, respectively, for all $2^4 + 2^8 = 272$ permutations. Every *active* UDW variable, i.e., $v_{\text{UDW}}^{2\text{CDW}} = 1$ or $v_{\text{UDW}}^{3\text{CDW}} = 1$, increases the sum by 1 and the current model is only valid if the overall weight is lower than the threshold. Due to the modeling of these capacity limits—the maximum number of CDWs—by the weighted constraints $\Phi_{\#\text{CDW}}$, the CDWs themselves have not to be modeled explicitly. This circumstance avoids a significant overhead concerning the number of Boolean variables and clauses.

5.2.2.1 Equivalence

One important property between uncompressed test data $\mathcal{I}$ and compressed test data $\mathcal{D}$ is the equivalence, which has to be fulfilled after $\mathcal{D}$ is decompressed on-chip (Lemma 4.8 on page 63). It is required that all bits in $\mathcal{I}$ are covered by exactly one replacement to ensure this equivalent: Uncovered as well as multiple covered bits violate this essential property. Consider the basic example in Table 5.3.

It has to be ensured that each bit position is covered by a segment (active). This can be achieved by adding one clause for each bit position as follows: Given the segment variables $v_1^{uC}, \ldots, v_s^{uC}$ covering bit position i, the corresponding clause would be

$$(v_1^{uC} + \cdots + v_s^{uC})$$

Equation 1 Φ_{uC} to ensure complete coverage of $\mathcal{I}$

$$\Phi_{\text{uC}} = (v_0^{1C} + v_0^{4C} + v_0^{8C}) \bullet (v_7^{1C} + v_4^{4C} + v_0^{8C})$$
$$\bullet(v_1^{1C} + v_0^{4C} + v_1^{4C} + v_0^{8C}) \bullet (v_2^{1C} + v_0^{4C} + v_1^{4C} + v_2^{4C} + v_0^{8C})$$
$$\bullet(v_3^{1C} + v_0^{4C} + v_1^{4C} + v_2^{4C} + v_3^{4C} + v_0^{8C})$$
$$\bullet(v_4^{1C} + v_1^{4C} + v_2^{4C} + v_3^{4C} + v_4^{4C} + v_0^{8C})$$
$$\bullet(v_5^{1C} + v_2^{4C} + v_3^{4C} + v_4^{4C} + v_0^{8C}) \bullet (v_6^{1C} + v_3^{4C} + v_4^{4C} + v_0^{8C})$$

Equation 2 Φ_{ME} to model mutual exclusion $\mathcal{I},\ ME = v_0^{4C}$

$$\Phi_{\text{ME}} = (\bar{v}_0^{4C} + \bar{v}_1^{4C}) \bullet (\bar{v}_0^{4C} + \bar{v}_2^{4C}) \bullet (\bar{v}_0^{4C} + \bar{v}_3^{4C})$$
$$\bullet(\bar{v}_0^{4C} + \bar{v}_0^{8C}) \bullet (\bar{v}_0^{4C} + \bar{v}_0^{1C}) \bullet (\bar{v}_0^{4C} + \bar{v}_1^{1C})$$
$$\bullet(\bar{v}_0^{4C} + \bar{v}_2^{1C}) \bullet (\bar{v}_0^{4C} + \bar{v}_3^{1C})$$

This clause is unsatisfied if none of these segment variables is activated, i.e., the segment is not covered at all. In this case, the complete problem instance is unsatisfied, which is due to the CNF. The conjunction of these clauses is represented by Φ_{uC}. The corresponding constraints for the example as stated in Table 5.3 are given in Equation 1. Furthermore, it has to be guaranteed that from the set of all segments, which cover the same bit position, only one segment is considered active. Since the mutual exclusive segments are known, implications are formulated as follows: Given a segment variable v^{uC} and the corresponding segment variables $v_1^{uC}, \ldots, v_t^{uC}$ of conflicting segments, the following clauses have to be added:

$$(\bar{v}^{uC} + \bar{v}_1^{uC}) \bullet \cdots \bullet (\bar{v}^{uC} + \bar{v}_t^{uC})$$

These clauses are unsatisfied when more than one segment variable is active from a conflicting segment set. This has to be formulated for each assigned segment variable. The conjunction of these clauses is denoted by Φ_{ME}. Table 5.4 shows the set of conflicting segment variables for the example to demonstrate the usage of mutual exclusion. Equation 2 gives the corresponding clauses for the mutual exclusions of the segment variable v_0^{4C}.

Retargeting Procedure
The constraints Φ_{RET} establish a link between the uncompressed test data $\mathcal{I}$ and the possible UDW permutations meaning the dictionary entries. Given is a segment of length u represented by the Boolean variable v_i^{uC}, hence, the segment covers a bit segment between position i and $(i-1)+u$. Since the specific assignment in $\mathcal{I}$ in this segment is known, the specific UDW w, which can cover this segment, is known as well. Therefore, the segment can only be active, if w is an entry in the dictionary, i.e., the variable $v_w^{2\text{CDW}}$ or $v_w^{3\text{CDW}}$ is assigned to "1." Since only a limited number of

Table 5.4 Mutual exclusions of segments (w/o single bit segments)

No.	Segment var.	Position [start:end]	Conflicting segment var.
1	v_0^{4C}	0:3	$v_1^{4C}, v_2^{4C}, v_3^{4C}, v_0^{8C}$
2	v_1^{4C}	1:4	$v_1^{4C}, v_2^{4C}, v_3^{4C}, v_0^{8C}$
3	v_2^{4C}	2:5	$v_1^{4C}, v_2^{4C}, v_3^{4C}, v_0^{8C}$
4	v_3^{4C}	3:6	$v_1^{4C}, v_2^{4C}, v_3^{4C}, v_0^{8C}$
5	v_4^{4C}	4:7	$v_1^{4C}, v_2^{4C}, v_3^{4C}, v_0^{8C}$
6	v_0^{8C}	0:7	All others

UDW variables can be assigned to "1" (due to $\Phi_{\#\text{CDW}}$), finding the most beneficial solution is the task of the solving algorithm.

For encoding these implications, the incoming test data $\mathcal{I}$ have to be analyzed during the problem instance generation. For every segment variable v_i^{uC}, the binary assignment between position i and $(i-1)+u$ as well as the corresponding specific UDW variables $v_{\text{UDW}}^{2\text{CDW}}$ and $v_{\text{UDW}}^{3\text{CDW}}$ have to be determined.
The clause

$$(\bar{v}_i^{uC} + v_{\text{UDW}}^{2\text{CDW}} + v_{\text{UDW}}^{3\text{CDW}})$$

is added to Φ_{RET} to make sure that if the segment variable is active, either $v_{\text{UDW}}^{2\text{CDW}}$ or $v_{\text{UDW}}^{3\text{CDW}}$ is assigned to "1," i.e., a dictionary entry exists for this segment. Since single bit segments are also encoded for every bit position, it is ensured that each incoming test data stream can be processed.

Example 5.1 Given the segment variable v_0^{4C}. The binary assignment in $\mathcal{I}$ at position [0 : 3] is assumed to be "0011." Then, the corresponding UDW variables are $v_{0011}^{2\text{CDW}}$ and $v_{0011}^{3\text{CDW}}$.

Various constraints have been generated and added to the overall PBO problem instance so far. In contrast to classic SAT, the proposed optimization SAT-based model aims at not just for a satisfying but for a solution of a certain quality. Due to this, an objective function is described in the following, which allows determining satisfying solutions of high quality and, hence, low costs.

5.2.3 *Optimization Function*

The constraints, which have been described so far, restrict the solution space of the problem instance such that all valid configurations are part of the solution space and invalid configurations are no solutions. However, an optimization function is strictly necessary to ensure the effectiveness of the approach. If the problem is formulated as a decision problem, each valid configuration can potentially be chosen. Most likely, a solving algorithm would choose a solution, in which each bit position is covered by a single bit codeword since this is a quite easy solution to be found.

Therefore, the quality of a configuration has to be encoded to guide the solving algorithm to find a cost-effective solution. The quality of a solution could be directly associated with the active segments in the data stream. However, for calculating the length of the compressed data stream accurately, three so- called cost-variables have to be associated with each segment. Implications on the active segments and used codewords can be used to determine the specific length of the compressed stream. However, preliminary experiments have shown that the reduction of active SBIs is most important for increasing the compression ratio. For improving the run-time, the optimization function was only used based on the knowledge of the segment length, i.e., using existing variables. The difference in the compression ratio—compared to the very accurate solution—is only marginal.

The optimization function $\mathcal{O}$ is formulated over all segment variables $v_1^{uC}, \ldots, v_n^{uC}$. The weight of the variables depends on the number of bits the segment covers. Obviously, the more bits a segment covers, the better the solution and, consequently, the smaller the weight.[1] The result of $\mathcal{O}$ for each solution is the sum of all weights, whose segment variables are active.

$$\mathcal{O}(\Phi_{\text{COST}}) = \sum_{i=1}^{n} m_i \cdot v_i^u \begin{cases} m_i = 4 & \text{if } u \equiv 1\text{C} \\ m_i = 2 & \text{if } u \equiv 4\text{C} \\ m_i = 1 & \text{if } u \equiv 8\text{C} \end{cases}$$

Besides the main optimization function $\mathcal{O}$, a secondary optimization function $\mathcal{O}_2$ is implemented. $\mathcal{O}_2$ is invoked during the optimization process to determine the length of the CDW that is to be used for an active UDW. This function determines whether a codeword of length 2 or 3 is assumed. The main idea is that UDWs, which occur more often, are encoded by shorter codewords with a length of 2. A counter $c_w^{\mathcal{D}}$ is introduced for every possible UDW w during instance generation to keep track of the number of occurrences in $\mathcal{D}$. $\mathcal{O}_2$ is built over all variables $v_w^{k\text{CDW}}$ with $k = \{2, 3\}$.

$$\mathcal{O}_2(\Phi_{\text{COST}}) = \sum_{j=1}^{272} 2 \cdot c_{w_j}^{\mathcal{D}} \cdot v_{w_j}^{2\text{CDW}} + 3 \cdot c_{w_j}^{\mathcal{D}} \cdot v_{w_j}^{3\text{CDW}}$$

Since $\mathcal{O}_2$ has a lower priority, it has no adverse impact on the result of $\mathcal{O}$, although, the length (here 2 or 3) of the CDW is determined for every active UDW such that the resulting costs are minimized. The run-time overhead of using $\mathcal{O}_2$ is negligible. Alternatively, this can be achieved in a post-processing step.

The PBO problem instance including both optimization functions are given to a PBO solver. Such a solver does not enumerate all solutions explicitly but uses

[1]Please note that the solving algorithms typically perform minimization.

learning techniques to traverse the search space effectively. During the search process, the best found solutions are enumerated until it is proven that no better solution exists. If the solving process aborts due to resource limits, the best solution found so far is returned as the satisfying model. A configuration and the corresponding sequence of CDWs, the compressed test data $\mathcal{D}$, are directly extracted from this satisfying model.

5.3 Optimization SAT-Based Retargeting Procedure

This section describes the optimization SAT-based retargeting algorithm, which has been seamlessly implemented within the existing heuristic retargeting framework of Sect. 4.3. This algorithm generates the PBO problem instance, whose components have been described in the last section in detail.

Algorithm 4 describes the retargeting procedure, which receives the incoming test data $\mathcal{I}$ and the hardware constraints concerning the maximal number of entries within the dictionary of length 2 by n_2 and of length 3 by n_3 as well as the set of supported UDW lengths u.

Algorithm 4 consists of the following parts:

1. The supported UDW lengths are individually processed to incorporate them in $\Phi_{\#\text{CDW}}$, in the constraint Φ_{COST} for the later optimization function and in the retargeting component Φ_{RET} itself. These steps are conducted for all lengths from lines 5 to 22.
2. The mutual exclusions have to be introduced into the PBO problem instance and the quality criteria have to be extended with the replacements. This is conducted from line 24 to 29.
3. As the last step, the overall PBO problem instance is concatenated and the objective function is initialized to consider the given hardware constraints. Thus, both weighted constraints ($\Phi_{\#\text{CDW}} \leq n_Y$) with $Y = \{2, 3\}$ and Φ_{COST} are appended, which models the necessary criteria for quality (lines 31 to 32). Furthermore, the PBO solver is called (line 33). After the determination of a satisfying model, different extraction procedure are invoked to a) generate the optimal configuration $\mathcal{C}$ (line 34) and the compressed test data $\mathcal{D}$ (line 35).

Step 1) above contains two subroutines, which are described in more detail in the following. The UDW permutations and the hardware constraints, which concern the maximum number of CDWs, have to be both configured in the embedded dictionary as follows:

(I) All possible UDW permutations for the supported UDW length are determined.

(II) Two Boolean variables $v_j^{2\text{CDW}}$ (line 8) and $v_j^{3\text{CDW}}$ (line 9) are allocated and the mapping between the UDWs and the corresponding variables are stored in the data container α in line 10.

(III) A weighted constraint $\Phi_{\#CDW}$ is built to enforce that less than n UDWs are active at the same time (lines 7 to 14).

Secondly, the linkage is established, which ensures that the corresponding UDW, depending on the data of $\mathcal{I}$, is configured if the segment is active. A further Boolean variable v_i^{uC} is allocated and the context of the variable is inserted in α. Depending on the UDW length, which is currently considered, the bit position j of the segment end is determined and is activated within a suitable replacement (lines 19 to 20).

Based on the developed PBO problem instance and the sketched formal retargeting procedure, the existing retargeting framework has been significantly extended. The advantages of introducing this optimization SAT-based approach are evaluated in the following by conducting different types of experiments.

5.4 Experimental Setup

This section describes the experimental setup of the proposed optimization SAT-based retargeting approach. The final results of the developed retargeting technique are distinguished against other existing techniques as well as the standardized IEEE 1149.1 protocol itself. Different test cases were considered for the experiments:

1. Random test data with sizes from 2048 (RTDR_2048) to 8192 (RTDR_8192) bytes (generated by a pseudo-random number generator based on Mersenne Twister).
2. *Golden* signature data for a JPEG encoder circuit given as input debug trace data following the technique of [DC14].
3. Commercially representative functional verification test cases of the MiBench benchmark suite [Gut+01], which were cross-compiled for a state-of-the-art softcore microprocessor by using an optimized library for embedded systems.

Two Verilog test benches TB_{LEG} and TB_{COMPR} are utilized for simulating the test data transfer to a CuT for validation of the obtained results. This setup is basically the same as the one described in Sect. 4.4 on page 69.

All retargeting procedures were executed on a Fedora 30 (with gcc 9.2.1) system holding an Intel Xenon E3-1240v2 3.4 GHz processor with 32 GB system memory. The implemented retargeting framework is written in C++ and clasp 3.1.4 is used as PBO solver [Geb+07].

5.5 Experimental Results

This section evaluates the conducted experiments concerning the high-entropic, functional verification test data of industrial representative circuits as well as an exemplary set of (debug) trace data.

Algorithm 4 Optimization SAT-based retargeting procedure

Require: test data $\mathcal{I}$, dictionary size n_2, n_3, UDW lengths u
1: Φ_{RET} := Empty PBO instance
2: $\Phi_{\text{\#CDW}}$:= Weighted constr. for max. entries of length 2 & 3
3: Φ_{COST} := Weighted clause for quality criteria
4: α := Map between meta-type and Boolean variable
5: **for all** $X \in u$ **do**
6: {Handling UDW permutations & max. entries}
7: **for** $i = 0$ **to** $i \leq 2^X$ **do**
8: $v_j^{2\text{CDW}}$ = allocBooleanVar() with j = DecToBin(i)
9: $v_j^{3\text{CDW}}$ = allocBooleanVar() with j = DecToBin(i)
10: $\alpha(UDW_j) \rightarrow (v_j^{2\text{CDW}}, v_j^{3\text{CDW}})$
11: $\Phi_{\text{\#2CDW}} = \Phi_{\text{\#2CDW}} + 1 \cdot \text{posLit}(v_j^{2\text{CDW}})$
12: $\Phi_{\text{\#3CDW}} = \Phi_{\text{\#3CDW}} + 1 \cdot \text{posLit}(v_j^{3\text{CDW}})$
13: $\Phi_{\text{COST}} = \Phi_{\text{COST}} + c_i \cdot \text{posLit}(v_j^{3\text{CDW}}) + c_i \cdot \text{posLit}(v_j^{2\text{CDW}})$
14: **end for**
15: {Handling replacements and link them to specific UDW}
16: **for** $i = 0$ **to** $i + X < \text{length}(\mathcal{I})$ **do**
17: v_i^{uC} = allocBooleanVar()
18: $\alpha(\text{uC}_i) \rightarrow v_i^{uC}$
19: $j = \mathcal{I}[i : (i-1) + X]$
20: $\Phi_{\text{RET}} = \Phi_{\text{RET}} \bullet (\text{negLit}(v_i^{uC}) + \text{posLit}(\alpha(UDW_j)))$
21: **end for**
22: **end for**
23: {Mutual exclusion of conflicting replacements & quality criteria}
24: **for all** $\text{uC}_\text{i} \in \alpha$ **do**
25: **for all** $\text{uC}_j \in$ conflictingOverlaps(uC_i) **do**
26: $\Phi_{\text{RET}} = \Phi_{\text{RET}} \bullet (\text{negLit}(\alpha(\text{uC}_i)) + \text{negLit}(\alpha(\text{uC}_j)))$
27: **end for**
28: $\Phi_{\text{COST}} = \Phi_{\text{COST}} + c_X \cdot \text{posLit}(\text{uC}_i)$
29: **end for**
30: {Performing search for model & invoking optimization process}
31: $\Phi = (\Phi_{\text{RET}} \bullet \Phi_{\text{COST}} \bullet (\Phi_{\text{2CDW}} \leq n_2)) \bullet (\Phi_{\text{3CDW}} \leq n_3))$
32: obj = $\mathcal{O}(\Phi_{\text{COST}})$ {objective function}
33: $\mathcal{M}$ = findSolution(Φ, obj)
34: $\mathcal{C}$ = ExtractConfigurationForActiveUDWs($\mathcal{M}, \alpha$)
35: $\mathcal{D}$ = ExtractReplacements($\mathcal{M}, \alpha$)
36: **return** $\mathcal{D}$, $\mathcal{C}$

Five different experiments are conducted for every benchmark as follows:

leg Invokes the IEEE 1149.1 protocol without any compression at all.

Huffman Implements a retargeting procedure by using a domain-specific Huffman algorithm, meaning that the existing hardware constraints concerning the TDR interface are considered internally.

heur Uses the greedy-like heuristic retargeting procedure as proposed by Sect. 4.3 while taking advantage of word-level operation to reduce the run-time.

opt Invokes the optimization SAT-based retargeting procedure using formal techniques with exhaustive traversal of the search space.

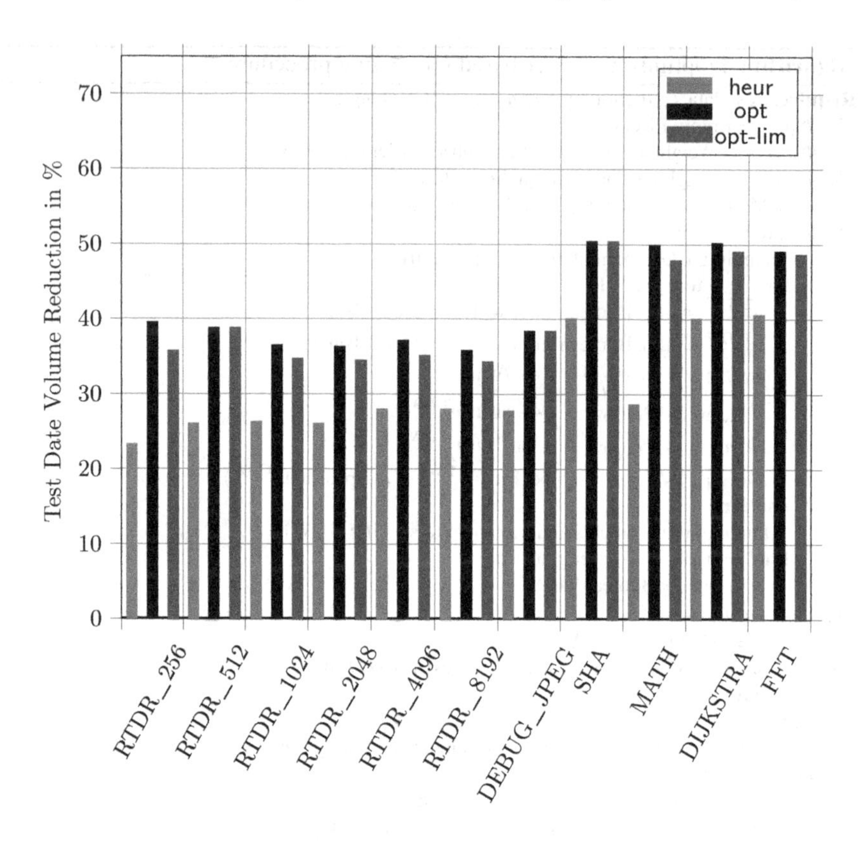

Fig. 5.2 Average TDV reduction for optimization SAT-based retargeting

opt-lim Implements a modified version of opt to limit the computational effort to reduce the resulting run-time significantly.

The results concerning the reduction of the TDV are given in detail in Table 5.6. The column size [bit] indicates the resulting data volume and column TDV red. [%] gives the data volume reduction ratio. Analogously, Table 5.7 presents the results concerning the number of test cycles (column #data cycles) and the overall TAT reduction (column TAT red. [%]). The TDV as well as TAT are plotted as histograms in Figs. 5.2 and 5.3 for a visual comparison.

All required configuration steps are considered in the measured data, i.e., they are included within the numbers of data cycles and data size. The consumed run-time in CPU minutes and the instance sizes, meaning the total number of introduced Boolean variables, are further presented in Table 5.5.

The results clearly demonstrate that the proposed opt technique is able to significantly improve both quality criteria—meaning TDV and TAT reduction—for all test cases compared to the legacy TAP, previously proposed techniques like [Ila+13] and also greedy-like heuristic approaches as proposed by Sect. 4.3

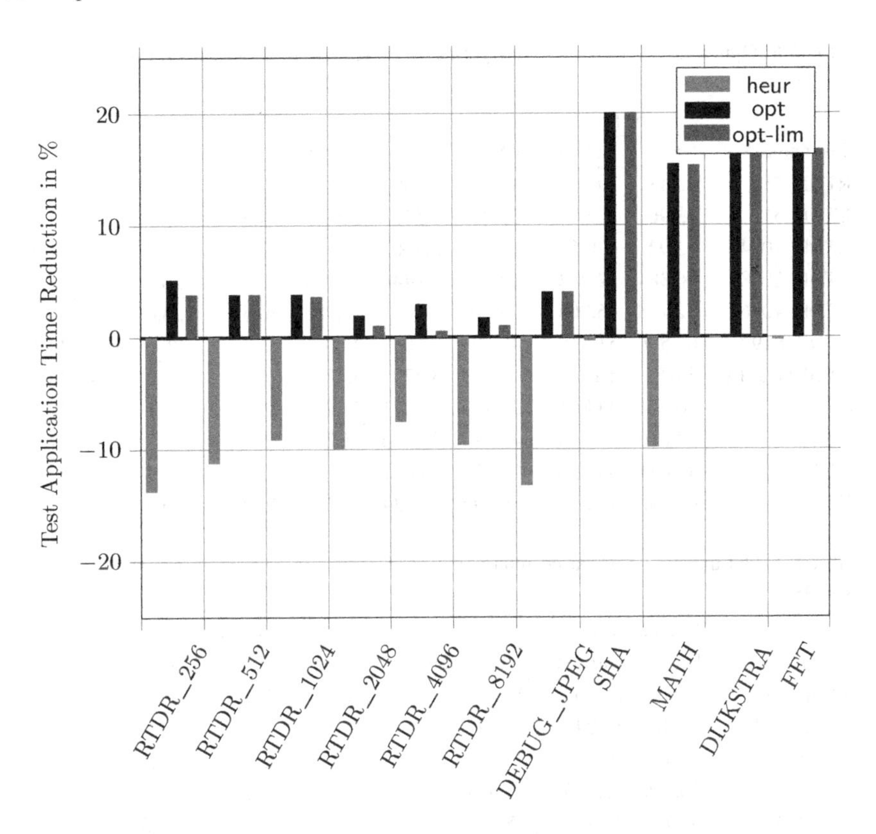

Fig. 5.3 Average TAT reduction for optimization SAT-based retargeting

Table 5.5 Run-time and instance sizes for retargeting random test data and debug data

	#Boolean variables	Run-time [min]			
Test name	opt/opt-lim	Huffman	heur	opt	opt-lim
RTDR_256	7,800	< 0.01	0.02	11.19	0.07
RTDR_512	15,059	0.01	0.04	21.14	1.69
RTDR_1024	29,678	0.03	0.07	43,75	0.39
RTDR_2048	58,930	0.05	0.19	97.59	0.76
RTDR_4096	117,203	0.10	0.63	190.89	14.85
RTDR_8192	233,997	0.22	2.16	492.08	29.47
DEBUG_JPEG	21,449	< 0.01	0.02	30.86	2.13
SHA	204,388	0.06	0.10	217.10	37.49
MATH	286,954	0.09	0.15	314.28	68.42
DIJKSTRA	222,101	0.06	0.10	183.78	38.76
FFT	289,219	0.08	0.14	356.07	72.11

Table 5.6 Optimization SAT-based retargeting: TDV reduction of random and industrial circuit designs

	Size [bit]					TDV red. [%]			
Test name	leg	Huffman	heur	opt	opt-lim	Huffman	heur	opt	opt-lim
RTDR_256	2,048	1,635	1,570	1,239	1,315	20.2	23.3	39.5	35.7
RTDR_512	4,096	3,298	3,027	2,516	2,516	19.5	26.1	38.8	38.8
RTDR_1024	8,192	6,465	6,035	5,200	5,350	21.1	26.3	36.5	34.7
RTDR_2048	16,384	12,973	12,109	10,444	10,739	20.8	26.1	36.3	34.5
RTDR_4096	32,768	26,005	23,607	20,626	21,282	20.6	28.0	37.1	35.1
RTDR_8192	65,536	51,119	48,529	42,068	43,057	22.0	26.0	35.8	34.3
DEBUG_JPEG	5,632	4,134	4,069	3,470	3,470	26.6	27.8	38.4	38.4
SHA	46,944	34,861	28,154	23,271	23,271	25.7	40.1	50.4	50.4
MATH	67,616	57,470	48,217	33,876	35,220	15.0	28.7	49.9	47.9
DIJKSTRA	49,441	34,056	29,172	24,621	25,165	31.2	40.1	50.2	49.1
FFT	66,880	48,856	39,743	34,034	34,310	26.9	40.6	49.1	48.7

Table 5.7 Optimization SAT-based retargeting: TAT reduction of random and industrial circuit designs

	#data cycles					TAT red. [%]			
Test name	leg	Huffman	heur	opt	opt-lim	Huffman	heur	opt	opt-lim
RTDR_256	2,053	2,580	2,334	1,949	1,973	−25.7	−13.7	5.1	3.8
RTDR_512	4,101	5,324	4,559	3,945	3,945	−29.8	−11.2	3.8	3.8
RTDR_1024	8,197	10,211	8,942	7,901	7,901	−24.6	−9.1	3.8	3.6
RTDR_2048	16,389	20,603	18,013	16,082	16,378	−25.7	−9.9	1.9	1.0
RTDR_4096	32,773	41,442	35,236	31,809	32,591	−26.5	−7.5	2.9	0.5
RTDR_8192	65,568	81,951	71,847	64,420	64,494	−25.0	−9.6	1.7	1.5
DEBUG_JPEG	5,637	6,871	6,125	5,412	5,412	−27.0	−13.2	4.0	4.0
SHA	46,949	51,921	47,090	37,561	37,561	−10.6	−0.3	20.0	20.0
MATH	67,621	84,851	74,253	57,152	57,264	−25.5	−9.8	15.4	15.3
DIJKSTRA	49,446	55,949	49,493	40,533	41,210	−13.2	−0.1	18.0	16.7
FFT	66,885	79,393	67,027	55,345	55,733	−18.7	−0.2	17.3	16.7

significantly. Compared to the legacy TAP, the TDV can be reduced for high-entropic test data on average by 36.4% and up to 37.1%. Previously proposed approaches achieved only an average reduction of 21.1% (Huffman) and of 26.7% (heur). The application on the debug trace data or on functional verification test data confirms these results. Here, the proposed technique is able to reduce the TDV by 47.6%.

When comparing the run-time of the different approaches, it has to be noted that the opt technique requires orders of magnitude more run-time than the heur technique, which is due to the hard computational task of solving a PBO problem instance. It is also shown that a large run-time benefit of this formal technique can be achieved even if the resources of the solving algorithm are limited (*opt-lim*), which decreases the compression ratio only negligibly. The opt-lim approach holds slightly different parameters for the PBO solver, which control, among others, the enumeration process. Furthermore, the shortcoming of the heuristic retargeting technique—as initially proposed in Sect. 4.3—concerning the TAT overhead when processing high-entropic test data has been completely eliminated. Approximately 26% more data cycles are needed when using the Huffman approach and the heuristic technique needs approximately 9% more data cycles on high-entropic test data, which was due to the fact that the dictionary entries were not well chosen due to the greedy-like manner. Consequently, a large number of SBIs had to be inserted in the resulting compressed test data $\mathcal{D}$. The proposed optimization SAT-based approach does not suffer from this circumstance due to the use of formal optimization techniques and of introducing an optimal configuration $\mathcal{C}$ for the embedded dictionary. It is able to even reduce the data cycles by 2.2% on average and up to 2.9% for random high-entropic test data and by 15% on average and up to 20.0% for the debug trace or functional verification data.

5.6 Summary and Outlook

This chapter proposed an optimization SAT-based approach for retargeting incoming test data using a codeword-based compression architecture for IEEE 1149.1-compliant TAP controller. The problem of finding optimal codewords is modeled as a PBO problem and formal solving techniques are leveraged to find beneficial codewords and, hence, an optimal configuration for the embedded dictionary.

Several experiments have shown that the TDV and the TAT reduction is significantly improved compared to other existing techniques by using the proposed formal techniques. In fact, this technique reduces the TDV for functional verification test data of industrial representative circuits by up to 50.4%. The TDV is even reduced for high-entropic test data and debug trace data by up to 37.1%. Furthermore, the TAT is also reduced by up to 20.0% compared to the TAT of a legacy transfer while processing those functional verification test data.

In conclusion, two completely different retargeting techniques have been recently considered: a heuristic as well as an optimization SAT-based technique. Generally, both techniques achieve a significant TDV reduction. The heuristic technique allows a very fast retargeting. However, a measurable overhead concerning the TAT is introduced, particularly, when processing high-entropic test data. In contrast to this, the formal technique avoids any increase of TAT even in case of high-entropic data processing but introduces a huge computational effort. Due to a very high run-time, the application of formal techniques on huge TDV is limited, which is most likely required to address the arising challenges for the test and debug of the next generation ICs.

Future work concerns the introduction of a multi-valued logic to the optimization SAT-based retargeting procedure. This will allow reflecting X-values, whose actual assignment can be freely determined. Incorporating this higher degree of freedom in the retargeting process will improve the compression efficacy even more.

Chapter 6
Reconfigurable TAP Controllers with Embedded Compression

The application of formal techniques within the retargeting procedure provides a powerful mechanism to determine the most beneficial—in the sense of TDV and TAT reduction—compressed test data $\mathcal{D}$ as well as a configuration $\mathcal{C}$. This approach works well for small and mid-sized test data volume, though, the maximum size of test data that can be processed is strictly limited. This is since the PBO problem instance scales non-linearly with the size of the test data and, hence, the size of the PBO problem instance basically explodes when processing large data. Even if well-engineered PBO solvers are available, the run-time and the memory consumption are completely unsustainable, which are required to solve the monolithic instance.

To tackle this shortcoming, this chapter proposes a partitioning scheme for the optimization SAT-based retarget procedure, which allows separating the monolithic PBO instance in smaller parts, the partitions. These partitions can then be processed individually while consuming only a negligible amount of run-time for the retargeting procedure. Furthermore, a partial reconfiguration scheme is developed, which allows reconfiguring the entries of the embedded dictionary selectively to save a considerable amount of configuration data.

The structure of this chapter is as follows: Sect. 6.1 motivates the introduction of the partitioning and reconfiguration scheme and describes the necessary extensions of the optimization SAT-based approach. These extensions address the computational hard problem of solving massive monolithic PBO instances and, hence, enable them to compress even large test data volume. This section also demonstrates an exemplary reconfiguration and discusses the selection of different model parameters. The experimental setup and the experimental evaluation of this newly proposed partition-based approach are given in Sects. 6.2 and 6.3, followed by a summary in Sect. 6.4.

S. Huhn, R. Drechsler, *Design for Testability, Debug and Reliability*,
https://doi.org/10.1007/978-3-030-69209-4_6

6.1 Partition-Based Retargeting Procedure Using Formal Techniques

Introducing a partitioning scheme to the formal optimization SAT-based retargeting procedure is about significantly reducing the required run-time and memory consumption to determine the compressed test data $\mathcal{D}$ as well as the configuration $\mathcal{C}$. This procedure aims at avoiding a strong adverse impact on the TAT, as typically caused by structural approaches, while introducing only feasible run-time for the determination.

Partitioning is a well-known approach to separate hard computational monolithic tasks into multiple sub-tasks that are easier to solve. Generally, this works fine if it is possible to divide the overall computational task into multiple sub-tasks, which can be processed independently. Directly applying such a partitioning on an optimization SAT-based procedure implies that typically only local optima are determined and, consequently, the local optima can strongly deviate from the global optimum. The overall test data are split into different chunks to project the given retargeting task to a partitioning scheme. Every chunk is then processed individually (sub-task), which sharply reduces the required computational effort. As already stated above, two critical challenges arise when partitioning is introduced:

1. The independence of the sub-tasks is not given. Thus, the data dependencies between these sub-tasks have to be considered by the formal model.
2. A strong deviation exists between the local and the global optimum—in the sense of highest TDV and TAT reduction—which will affect the effectiveness of the retargeting procedure. To avoid an adverse impact on global effectiveness, every local optimum must be reached, implying that the local and the global optimum converge.

Locally means that only one chunk (sub-task) of the overall test data is processed, hence, a determined configuration works best only for the specific chunk. Analogously, global refers to a configuration that is determined by considering all test data at-once and works best globally. However, this calculation requires a very high run-time for large TDV. To address this deviation, it is targeted to accomplish all local optima by reconfiguring the dictionary for every chunk. However, this introduces significantly more configuration data. This data overhead is tackled by introducing a partial reconfiguration scheme, which is aware of the current state of the dictionary and reconfigures entries only if beneficial.

Definition 6.1 Let $\mathcal{I}$ be a sequence of bits $(b_1, b_2, .., b_N)$, which represents the target data for partitioning and let $\#\mathcal{I} = N$ be the number of bits in the sequence. Then, a partition $P_{i,j}$ is defined as a coherent subsequence $(b_i, b_{i+1}, b_{i+2}, \ldots, b_j)$ with $i \leq j$ and $i \geq 1$ and $j \leq N$. The size of partition $P_{i,j}$ is defined by $\#P_{i,j} = j - i + 1$.

6.1.1 Formal Partitioning Scheme

Partitioning aims at splitting the overall incoming test data $\mathcal{I}$ into separated parts, as formally stated in Theorem 6.1. Every part is then retargeted individually, which allows reducing the run-time significantly. Multiple aspects have to be considered when introducing a partition scheme on top of the existing retargeting procedure. The selection of suitable partitions is a crucial aspect since uncovered interdependencies have to be strictly avoided. Otherwise, the correctness of the retargeting procedure cannot be ensured. In particular, the equivalence has to be fulfilled between the UDW and the CDW after decompression on-chip. For achieving equivalence of the retargeted data when introducing partitions, the selection has to ensure that no bit position is covered more than once by a partition, as formally stated in Theorem 6.2.

Lemma 6.2 *Let $P_{i,j}$ and $P'_{i',j'}$ be two partitions of $\mathcal{I}$.*
Then, the partitions P and P' are free of intersecting bits meaning that $j < i'$ or $j' < i$ is valid. With $P_{i,j} \cap P'_{i',j'} = \varnothing$, no bit position in $\mathcal{I}$ is included in more than exactly one partition.

Even if the partition is deliberately selected, the determined C is just optimal for a single partition (local optimum). Preliminary experiments have shown that if one set of locally optimal codewords is applied to the complete data stream, the effectiveness of the approach is reduced, which leads to only a low reduction of the TDV and the TAT compared to the achieved results when globally optimal codewords are applied.

Multiple Configurations
Multiple configurations are determined to tackle the problem of codewords being only locally optimal. In particular, one configuration is determined for every processed partition. This new configuration adapts the embedded dictionary individually to the partition following the scheme as shown in Fig. 5.1 on page 79. Here, the embedded dictionary is configured by C_i before the specific compressed test data chunk c_i is transferred to the CuT, which implements the compression-based TAP controller. This reconfiguration solves the problem concerning local and global optima well. However, this introduces large configuration data, which leads to additional data volume as well as cycles. The maximum number of codewords, which are reconfigured during a reconfiguration, can be limited to control the introduced overhead by multiple configurations.

Partial Reconfiguration
The reconfiguration of the dictionary is only applied partially to avoid an adverse impact on the TDV and, particularly, on the TAT. Thus, the retargeting procedure considers the current state S_D of the dictionary, i.e., the codewords, which have been configured while processing the previous partition. It is aimed at *reusing* already configured entries for the following partition to, eventually, reduce the overall configuration data as stated in Theorem 6.3.

Lemma 6.3 *Given is an ordered sequence of partitions* $(P_1, P_2, \ldots, P_m)$ *and the current state* $S_D@0$ *of the dictionary at point of time* $t = 0$*, i.e., the codewords for all entries within. The point of time* $t = 0$ *represents the initial state of the dictionary (due to the synthesis of the TAP controller).*
Then, the retargeting procedure ρ *receives a partition* P_i *and the current state* $S_D@i-1$*, which is determined due to previous configurations* C_j *with* $0 \leq j \leq i-1$*. Furthermore, a configuration* C_i *is partial if only a subset of entries is included.*

6.1.2 *Exemplary Reconfiguration*

The partial reconfiguration scheme is demonstrated in Table 6.1, which shows exemplary data for multiple reconfigurations of the embedded dictionary by applying C_0 to C_n. As stated, every C_i was determined by ρ that processes the partition P_i (representing a chunk of the $\mathcal{I}$) with respect to the current state of the dictionary. Column No. shows the number of the dictionary entry, column C_0 represents the default state of the dictionary, i.e., the default configuration which is programmed due to synthesis. It is assumed that the dictionary holds 12 dynamically configurable entries, which can contain half-byte or byte-long entries.[1] The columns C_1 up to C_n represent the state **after** applying the configuration C_i. The bottom line $\sum C$ gives the size of the current configuration in bit. In the case of C_{n-1}, only entries 1 to 8 are included. Thus, entries 9 to 12 remain in the previous state. C_n further configures the dictionary partially, particularly, the entries 11 and 12 are not reconfigured either. Both entries remain in the state, in which they have been set several configuration phases ago.

This example clearly demonstrates that the size of the configuration data directly 130 scales with the number of included entries as well as with the length (half-byte 131 or byte-long) of the indvidual entry. The resulting configuration size varies from 79 bit (C_1) to 36 bit (C_n). Consequently, it is necessary to consider the resulting size of the configuration data while applying the partitioning scheme.

6.1.3 *Parametric Analysis*

This section briefly introduces the required adoption to incorporate the partitioning scheme for the optimization SAT-based retargeting procedure and describes the analysis for determining the most suitable set of parameters.

The first partition P_0 is processed by a PBO instance, which follows the scheme as presented in Sect. 5.3 on page 85. All clauses and variable assignments have

[1] For a fair comparison, exactly the same parameters (hardware constraints) are assumed as used for the heuristic as well as optimization SAT-based retargeting procedures.

Table 6.1 Partial Reconfigurations of the embedded dictionary with $\mathcal{C}_0$ to $\mathcal{C}_n$

No.	$\mathcal{C}_0$	$\mathcal{C}_1$	...	$\mathcal{C}_{n-1}$	$\mathcal{C}_n$
1	1111	01011010		1010	0111
2	0101	1110		00010110	0000
3	0110	0101		0110	11110001
4	00110011	10010110		01111000	00110010
5	01010101	1100		0011	$\mathcal{C}_{n-1}$
6	1010	0101		00011111	01010101
7	0000	10001011		0111	$\mathcal{C}_{n-1}$
8	10101010	11101010		1010	$\mathcal{C}_{n-1}$
9	1000	10001111		$\mathcal{C}_{n-2}$	$\mathcal{C}_{n-2}$
10	1001	10010100		$\mathcal{C}_{n-2}$	1000
11	0001	0100		$\mathcal{C}_{n-2}$	$\mathcal{C}_{n-2}$
12	11001100	11101111		$\mathcal{C}_{n-2}$	$\mathcal{C}_{n-2}$
$\sum \mathcal{C}$ [bit]	–[a]	79		44	36

[a] Default configuration due to synthesis

to be removed from the PBO problem instance, which refer to the $\mathcal{I}$ chunk $(u_i, u_{i+1}, \ldots, u_j)$, to retarget any succeeding partition $P_{i,j}$. Furthermore, the configuration state $\mathcal{S}$ of the dictionary is extracted after every retargeting step since this configuration state has to be considered for any subsequent one. A lookup function σ is implemented, which includes the current configuration state dictionary $\mathcal{S}_D$, receives a compressed data word cdw_i, and checks if cdw_i is included in the current configuration of the dictionary such that

$$\sigma(cdw_i) = \begin{cases} 1 & \text{if } cdw_i \in \mathcal{S}_D \\ 0 & \text{else} \end{cases}$$

The optimization function has to be modified such that the configuration of an already configured compressed dataword cdw_i ($\sigma(cdw_i) = 1$) does not introduce any further costs in the sense of configuration data. Furthermore, two primary parameters have to be deliberately selected since both strongly influence the size of the configuration data as well as the consumed run-time for the retargeting procedure. The effect of the following two parameters have been extensively elaborated for random test data as follows:

Maximal Number of Codewords in Reconfiguration r_s

The overall size of the configuration $\mathcal{C}_i$ directly scales with the number of codewords (included in $\mathcal{C}_i$) as demonstrated in Table 6.1. This number cannot exceed the overall number of dynamically configurable entries. Reconfiguration ratios between a full reconfiguration (100%), a half-reconfiguration (50%), and the two intermediate ratios 66.7% and 83.3% were investigated. Note that these shares represent the

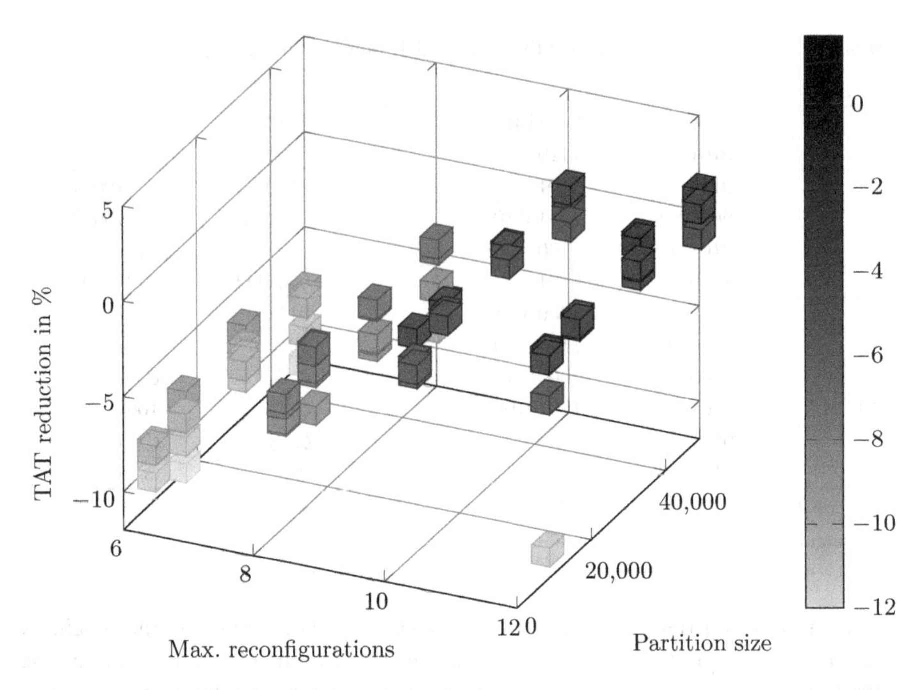

Fig. 6.1 Parameter identification considering TAT

maximal number of allowed codewords reconfigurations, which does not necessarily mean that the number of entries is actually reconfigured.

Maximal Partition Size p_s
The maximum size of a partition p_s controls the computational effort, which is required to process this partition. This effort, i.e., the size of the PBO instance, scales non-linearly with the size of the input data for the retargeting procedures. Different partition sizes of 8K, 16K, 32K, and 48K were investigated.

Figure 6.1 shows the parameter study for the investigated $p_s \in \{8K, 16K, 32K, 48K\}$ at the y-axis and the maximum reconfigurations $r_s \in \{6, 8, 10, 12\}$ at the x-axis while processing high-entropic test data with sizes from 64K to 1,024K. Each experimental run is represented by a data point and the resulting TAT reduction in percentage is plotted at the z-axis. Besides this, the TDV reduction is stable with an average ratio of 37.3% and a variance of 2.3%. However, the TAT varies between a slight reduction but also between a slight increase (compared to non-compressing data transfer). Hence, the TAT reduction in % is represented by the gray coloring scheme: the darker the gray, the higher the TAT reduction and vice versa.

The impact of the parameters on the resulting run-time is given in Fig. 6.2. Here, the $p_s \in \{8K, 16K, 32K, 48K\}$ are shown at the y-axis and the maximum reconfigurations $r_s \in \{6, 8, 10, 12\}$ at the x-axis while processing high-entropic test data with sizes from 64K to 1,024K. Each experimental run is represented by a data point and the normalized run-time in seconds is plotted at the z-axis. The

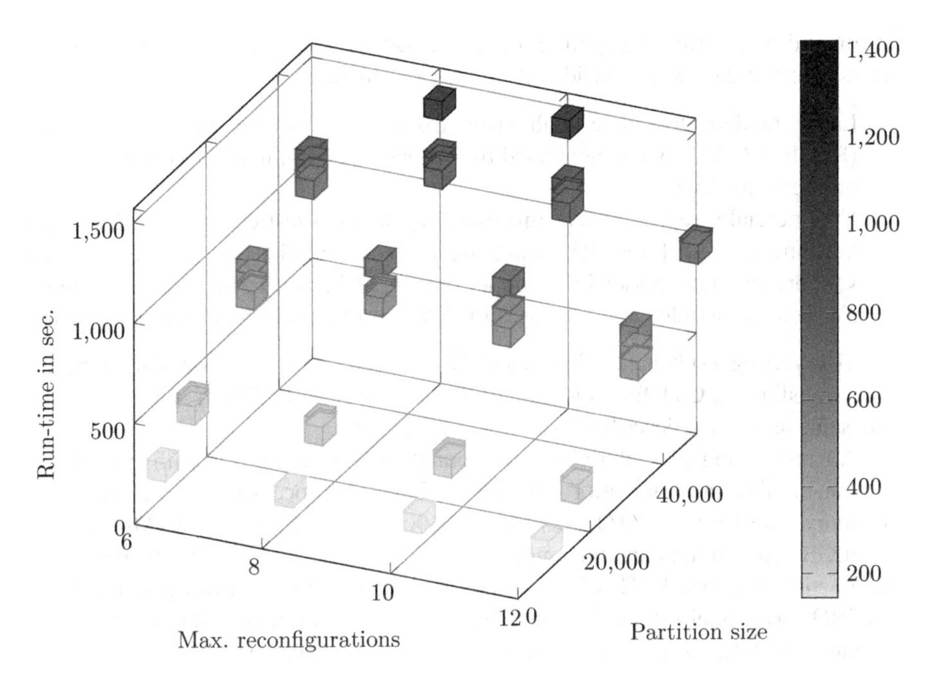

Fig. 6.2 Parameter identification considering run-time

normalization concerns the incoming test data and uses a size of 64k as a reference. The same gray coloring scheme is used in this Figure: the darker the gray, the higher the resulting run-time and vice versa.

As shown by Figs. 6.1 and 6.2, determining a suitable parameter set implies a trade-off between a reduced TAT or a reduced run-time. The results show that the parameter set of 12 reconfigurations and a partition size of 32,768 bits represents a fair compromise. Further experiments were conducted, which have shown that this set of parameters is suitable for non-random test data. This is due to the fact that random test data are typically harder to compress since the entropy is higher and, thus, compressing random data forms the worst case in the sense of compression [BT07].

6.2 Experimental Setup

This section describes the experimental setup of the proposed partition-based retargeting approach, which incorporates the optimization SAT-based retargeting procedure of Chap. 5.

The final results of the developed partition-based retargeting procedure part are distinguished against the legacy transfer by using JTAG leg, the heuristic approach heur, and both formal retargeting procedures opt as well as opt-lim. The following classes of test data were considered for the experiments:

1. Large random test data with sizes from 8,192 (RTDR_8192) to 131,072 (RTDR_131072) bytes (generated by a pseudo-random number generator based on Mersenne Twister).
2. Commercially representative functional verification test cases of the MiBench benchmark suite [Gut+01], which were cross-compiled for a state-of-the-art softcore microprocessor by using an optimized library for embedded systems. These tests include large test cases for JPEG compressors and decompressors.

Two Verilog test benches TB_{LEG} and TB_{COMPR} are utilized for simulating the test data transfer to a CuT for validation of the obtained results. This setup is basically the same as the one described in Sect. 4.4 on page 69.

All retargeting procedures were executed on a Fedora 30 (with gcc 9.2.1) system holding an Intel Xenon E3-1240v2 3.4 GHz processor with 32 GB system memory. The *Time-Out* (TO) was set to 86.400s and the *Memory-Out* (MO) was set to 30GB. The original and uncompressed test data was processed by the developed retargeting framework. This framework is written in C++ and invokes clasp 3.1.4 as PBO solver [Geb+07] and, eventually, generates the compressed test data as well as the configuration for the embedded compression architecture.

6.3 Experimental Results

This section evaluates the conducted experiments of processing large test data volumes with up to 1,000K of bits. In Table 6.2, the results concerning the TDV reduction of the conducted experiments are presented as follows: The column size [bit] indicates the resulting data volume and column data reduction [%] gives the data volume reduction ration. Analogously, Table 6.3 shows the results concerning the number of test cycles (column #data cycles) and the overall TAT reduction (column cycle reduction [%]). Here, heur refers to the heuristic retargeting procedures of Sect. 4.3 and opt as well as opt-lim to the previously introduced optimization-based techniques. The column part presents the resulting TDV and TAT reduction while using the proposed partitioning scheme with p_s = 16K and (up to) a full-reconfiguration including the reconfiguration data and cycle overhead.

The TDV of the part technique is robust and comparable to opt-lim as shown in Table 6.2. For instance, part allows compressing 1,024K high-entropic test data by 37.3% and functional verification test data by up to 62.8%. The experiments generally show that the resulting TDV reduction is lower for high-entropic than for functional verification test data. Concerning the TAT reduction for the most critical high-entropic test data, the resulting TAT reduction of part is slightly lower compared to opt-lim while processing the hard-to-compress high-entropic data. However, the ratios are significantly less compared to using heur, though, the ratios

Table 6.2 Optimization SAT-based retargeting with partitioning: TDV reduction of random & industrial circuit designs

	Size [bit]					TDV red. [%]			
Test name	leg	heur	opt	opt-lim	part	heur	opt	opt-lim	part
RTDR_8192	65,536	48,529	42,068	43,057	42,575	26.0	35.8	34.3	35.0
RTDR_16384	131,072	89,952	TO	83,476	84,312	31.4	TO	36.3	35.7
RTDR_32768	262,144	177,991	TO	166,299	164,700	32.1	TO	36.6	37.2
RTDR_65536	524,288	355,682	TO	MO	327,448	32.2	TO	MO	37.5
RTDR_131072	1,048,576	711,858	TO	MO	657,303	32.1	TO	MO	37.3
Patricia	76,864	29,827	31,678	31,781	28,807	61.2	58.7	58.7	62.5
sort	76,832	27,444	32,863	26,546	28,585	64.3	57.2	65.4	62.8
Dijkstra	77,441	32,265	33,673	32,793	32,594	58.3	56.5	57.7	57.9
bmath	109,793	48,825	TO	50,013	46,545	55.5	TO	54.4	57.6
blowfish	143,232	73,579	TO	71,242	67,442	48.6	TO	50.3	52.9
cjpeg	703,648	423,315	TO	TO	372,719	39.8	TO	TO	47.0
djpeg	801,922	481,904	TO	TO	417,133	39.9	TO	TO	48.0

Table 6.3 Optimization SAT-based retargeting with partitioning: TAT reduction of random & industrial circuit designs

	#data cycles					TAT red. [%]			
Test name	leg	heur	opt	opt-lim	part	heur	opt	opt-lim	part
RTDR_8192	65,541	71,847	64,420	64,494	65,744	−9.6	1.7	1.5	−0.3
RTDR_16384	131,077	151,351	TO	128,427	130,160	−15.5	TO	2.0	0.7
RTDR_32768	262,149	300,151	TO	257,168	257,912	−14.5	TO	1.9	1.6
RTDR_65536	524,283	600,532	TO	MO	517,048	−14.5	TO	MO	1.4
RTDR_131072	1,048,581	1,201,183	TO	MO	1,046,149	−14.6	TO	MO	0.2
Patricia	76,869	51,622	54,870	54,926	50,705	32.8	28.6	28.5	34.0
sort	76,837	47,733	45,002	45,051	54,074	37.9	41.4	41.4	29.6
Dijkstra	77,446	55,097	59,631	54,945	55,941	28.9	23.0	29.1	27.8
bmath	109,798	82,875	TO	80,839	85,680	24.5	TO	26.4	22.0
blowfish	143,237	126,427	TO	55,426	114,233	11.7	TO	61.3	20.3
cjpeg	703,653	715,545	TO	TO	626,735	−1.7	TO	TO	10.9
djpeg	801,927	803,661	TO	TO	707,152	−0.2	TO	TO	11.8

are lower compared to opt-lim while applying the proposed technique on functional verification data. In the case of processing 64K, heur causes a TAT increase of 14.5% while part achieves a TAT reduction of 1.6%. The TDV and the TAT are further plotted in Figs. 6.3 and 6.4 for a visual comparison.

Figure 6.5 presents a run-time comparison (logscale on y-axis) for different test data volumes of both formal techniques (opt and opt-lim) versus the proposed part technique. As clearly shown, the run-time of the proposed approach scales with increasing data volume in a feasible way. For instance, part consumes 119s to process 16K and 9,987s to process 1,024K, i.e., the scale factor of run-time

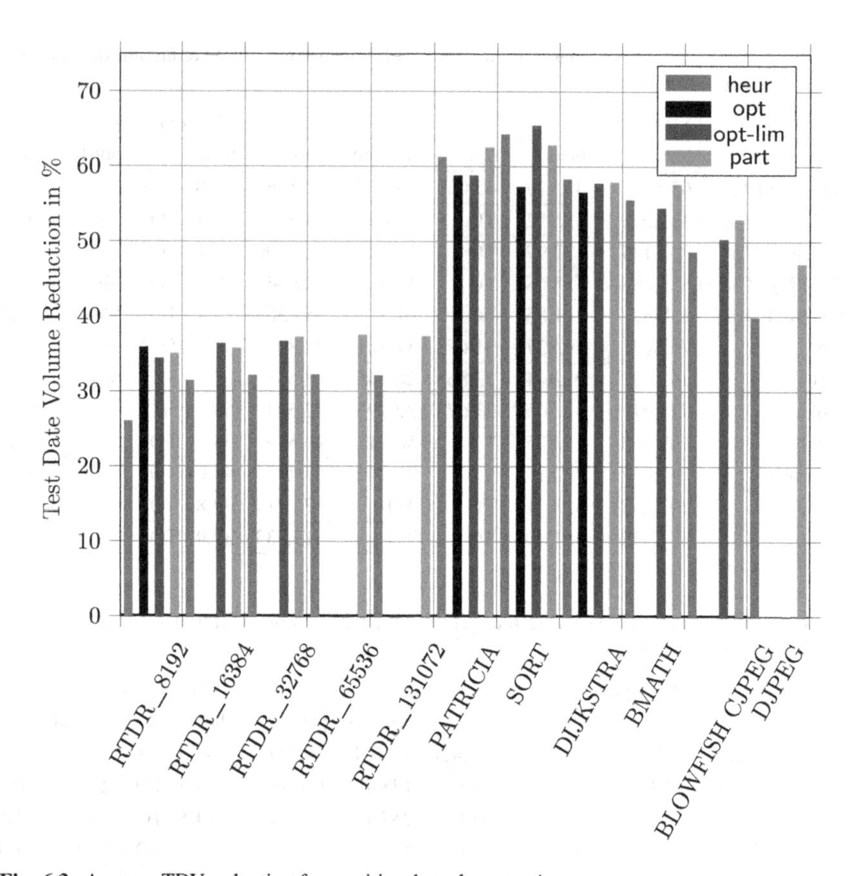

Fig. 6.3 Average TDV reduction for partition-based retargeting

against volume increase is $\frac{83.9}{64} \approx 1.31$. In comparison to this, opt-lim scales with $\frac{279.9}{16} \approx 17.49$ while processing 16K to 256K before exceeding the TO. opt has already exceeded the TO when processing 128K of test data.

6.4 Summary

This chapter proposed a new formal optimization-based retargeting technique to process even large and high-entropic test data such that the advantages of compression-based TAP controllers can be leveraged. The proposed retargeting technique incorporates a partition-scheme to reduce the resulting search space of the underlying formal model such that the computation terminates in a reasonable time. More precisely, the basic embedded-dictionary principle of DDU being introduced by VecTHOR is enhanced by a partial reconfiguration capability to significantly reduce the reconfiguration overhead.

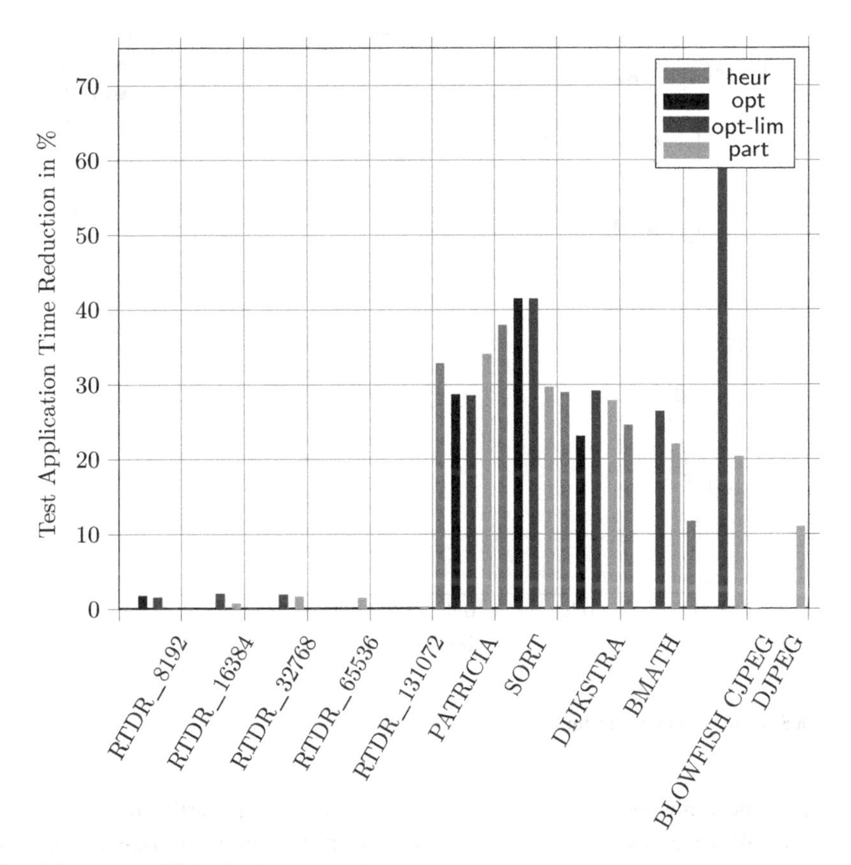

Fig. 6.4 Average TAT reduction for partition-based retargeting

Experiments have shown that the run-time was significantly reduced compared to other existing formal techniques. In particular, the scaling factor of run-time vs. TDV was reduced by magnitudes from 17.49 to 1.31 while retaining the TDV and the TAT reduction ratios. Due to the reduced run-time, random test data with more than one million of bit were successfully retargeted while reducing the TDV by approx. 37%. Even more importantly, the TDV of functional verification test data of industrial representative circuits were reduced by up to 50.4%. In every test case considering larger test data volume, a further TAT reduction was achieved by up to approx. 30% in the case of test data of industrial representative candidates.

It has been clearly demonstrated that the optimization SAT-based retargeting procedure outperforms heuristic approaches since an optimal configuration for the embedded dictionary is determined. Due to the hard computational task of solving the PBO problem instance, this approach is more suitable for small and mid-sized test volumes. The run-time scales non-linearly when processing large test data volume as a monolithic PBO problem instance and, hence, is unsustainable. However, the introduction of a partitioning approach, which considers the current

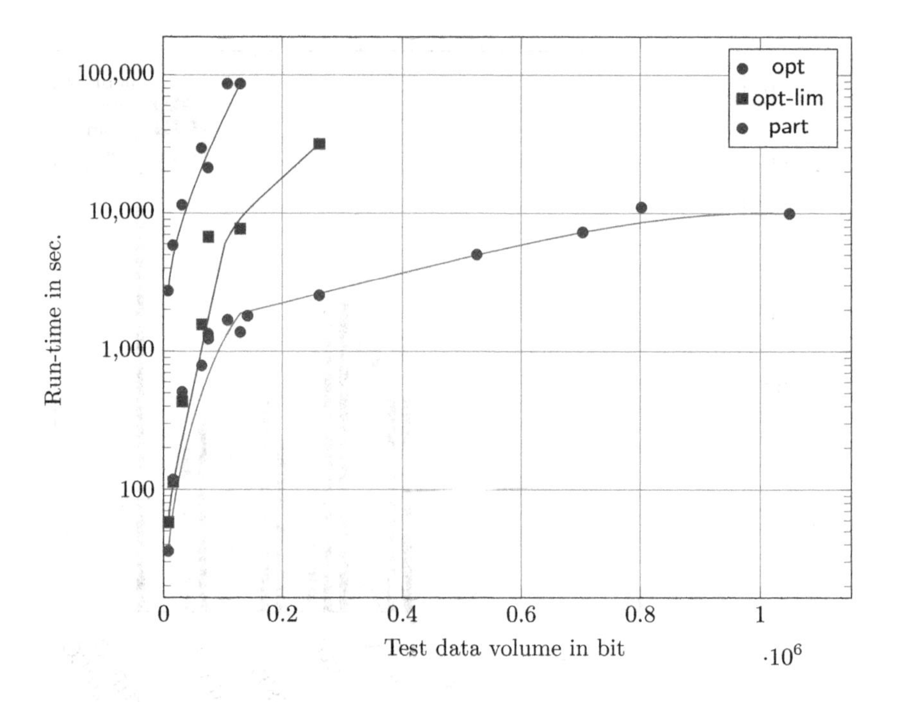

Fig. 6.5 Comparison of run-time of retargeting techniques

state of the embedded dictionary, in combination with a partial reconfiguration scheme, allows unleashing the full potential of optimization SAT-based technique. The proposed approach yields a significant TDV and TAT reduction, even though more than one million of bits are considered.

Chapter 7
Embedded Multichannel Test Compression for Low-Pin Count Test

The latest accomplishments in the field of design and manufacturing of ICs enable entirely new application scenarios. For instance, the newest generation of electronic control units integrates a large number of sophisticated onboard ICs to implement advanced driver-assistance systems. Due to the safety-critical aspect of such an application, different types of regulations, standards as well as practices have to be considered. Potential defects, which may have been occurred during the manufacturing process, have to be reliably detected to be compliant with these challenging requirements [WW03]. Thus, a very high test coverage is required leading to large sets of test patterns, which have to be executed during the manufacturing test [MR16].

One further challenge concerns the available pin count. In industrial practice, the test is applied during different test insertions. The available infrastructure may vary significantly in different environments. For instance, the burn-in testing is an important step to ensure the required quality of the manufactured IC, which is performed under stressing environmental conditions like an elevated temperature. Even when these harsh conditions allow accelerating the detection of infant mortalities, i.e., the latent early-life failures, this type of test still forms a bottleneck of the overall testing [KS03]. Furthermore, dedicated burn-in devices are required, which provide only a very limited number of test pins. Other test insertions use massive parallelization to reduce test costs, which again causes restrictions in terms of pin count.

Complex designs utilize powerful embedded test compression techniques like [JT98, JGT99, ICM99, ZL77, Raj+04, LC03, CC01] as already discussed in more detail in Sect. 4.1. These techniques aim at reducing the overall testing time and, thus, saving costs. Typically, dedicated hardware is embedded on-chip that allows an on-the-fly decompression of the (compressed) test data during the transfer between the ATE and the CuT.

Commercially available state-of-the-art compression techniques like [Raj+04] achieve a significant compression ratio by taking advantage of certain structural

S. Huhn, R. Drechsler, *Design for Testability, Debug and Reliability*,
https://doi.org/10.1007/978-3-030-69209-4_7

properties of the test data. The compression ratio of these techniques scales with the share of X-values and depends on further parameters like the number of inserted scan chains. In general, either a large number of X-values has to be included in the test data or the generation of test cubes has to be considered in the compression procedure [MK06] forming a run-time intensive task. This works fine for many of the test patterns, however, the remainder of the test patterns cannot be compressed at all; they are rejected due to the introduced compression architecture. Techniques like [Ace+17] introduce conflict-aware test points to further increase the compacting abilities of the test set, even though this does not resolve the rejection of a single test pattern.

Generally, the test pattern has to satisfy a large number of constraints coming from the compression architecture in order to be compressible. Since this is not always possible, these rejected tests lead to fault coverage loss. This effect is highly amplified in LPCT environments, which have been described above. In such cases, the number of incompressible patterns increases significantly. This is since state-of-the-art compression solutions like [Dho+16, LR12] have the inherent characteristic that the compression efficacy is low if there are only very few pins, and therefore data channels, available.

Every *Rejected Test Pattern* (RTP) is crucial since a simple neglection would decrease the test coverage and, hence, violate the test quality requirements. Thus, these RTPs have to be applied during the test to avoid any loss of test coverage. These RTPs are then transferred to the chip in a sequential and completely uncompressed fashion, i.e., bypassing the compression architecture. This yields to an adverse impact on the overall compression ratio and a significant test cost increase.

This chapter proposes a novel hybrid architecture, which seamlessly combines a state-of-the-art embedded test compression technique with a codeword-based compression scheme replacing the costly bypass structure. This approach has been developed in a tight cooperation with Infineon Germany to, among others, ensure that industrial relevant scenarios are considered and, furthermore, correct assumptions from industrial perspective are made. The codeword-based approach is meant to be applied for the RTPs to tackle the shortcomings of existing techniques. More precisely, the adverse impact of RTPs on the overall compression ratio is significantly reduced. The introduced hardware overhead is negligible and a powerful retargeting approach has been implemented based on the optimization SAT-based retargeting approach as described in Chap. 5. The proposed architecture also allows the utilization of existing multichannel test compressions structures, which are typically introduced in industrial-sized designs. This allows reducing the transfer time of the RTPs even more and, hence, the resulting TAT. Experiments have shown that a TDV reduction of up to 67.4% for the RTPs can be achieved. The resulting TAT can be further reduced by up to 64.7% when orchestrating existing multichannel capabilities.

The structure of this chapter is organized as follows: Sect. 7.1 briefly describes related works concerning embedded test compression techniques including further approaches based on codewords. Section 7.2 draws the proposed hybrid compression architecture, which mainly consists of three different components being successively described in the subsections. Extended topics are presented in Sect. 7.3, which concern a hardware cost analysis and an extension which takes advantage of an existing multichannel test topology to further optimize the compression efficacy. The experimental setup is described in Sect. 7.4 and the conducted results are presented in Sect. 7.5. Finally, Sect. 7.6 summarizes the paper.

7.1 Related Works

Within the last decade, many different test compression techniques have been proposed in the literature, which all aim at reducing TDV and TAT. The TDV scales directly with the required memory resources on the ATE, which is one strictly limited resource, in particular when considering test cost reduction techniques like multi-site testing [Iye+02].

Architecture of Embedded Test Compression

It is required to embed dedicated hardware on-chip to enable the transfer of compressed test data $\mathcal{D}$ from the (external) test equipment to the CuT to take advantage of an embedded test compression technique. In Fig. 7.1, the principle architecture is shown as follows: The newly inserted hardware implements, among others, an on-the-fly decompressor, which allows decompressing $\mathcal{D}$ on-chip without any loss of information. The uncompressed test data $\mathcal{I}$, meaning the test patterns, are determined by ATPG tools. These data are then post-processed by a retargeting procedure applying the actual compression scheme results in the compressed test data $\mathcal{D}$.

The state-of-the-art techniques like [Raj+04] achieve a very high compression ratio, which mostly scales with the share of X-values in the processed test patterns. Thus, this compression ratio varies strongly depending on the structural properties of the data, e.g., most compression architectures apply constraints on the distribution of scan cell assignments. A more critical circumstance is that a certain share of the overall pattern set is incompressible at all since the generated test patterns do not satisfy the constraints of the compression architecture. It is not possible that they are decompressed on-chip without any loss of information, which is due to the structural properties of these patterns. Thus, they are rejected by the compression architecture. These RTPs have then to be transferred via a dedicated bypass controller in an uncompressed and sequential fashion, which bypasses the decompressor-infrastructure completely. This results in a significant overhead regarding the TAT as well as the TDV. However, these RTPs have to be executed to avoid any loss in test coverage, which is most important in domains in which a zero defect policy is required.

Four components have to be implemented on-chip to instantiate such an embedded compression technique. These components are as follows:

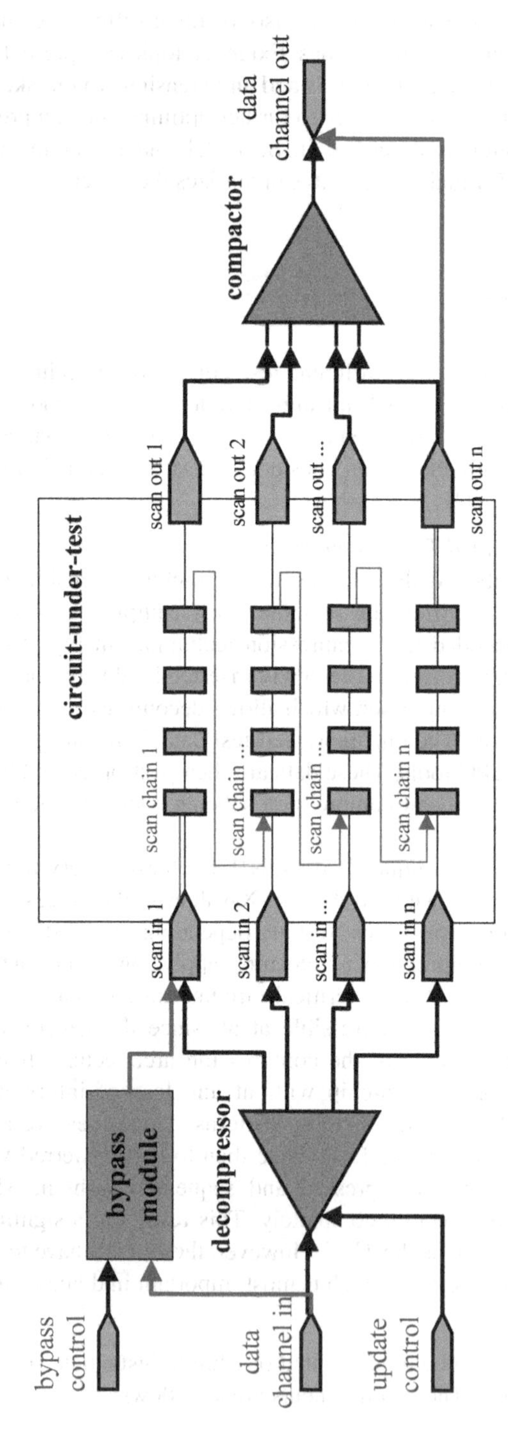

Fig. 7.1 Embedded test compression scheme [Raj+04]

1. The decompressor restores the uncompressed test patterns, which are then loaded into the scan chains,
2. the compactor determines the test responses of the circuit by processing the scan chains' outputs,
3. the bypass controller redefines the datapath (including tail-gating all scan chains) if the bypass is enabled for transferring uncompressed data (marked in red), and
4. a set of chip-level pins holding a clock signal, two control signals (for the compactor/decompressor and the bypass controller) as well as data input and a data output channel for the compressed test pattern and test response.

Typically, multiple data channels are introduced when considering industrial-sized designs. This allows extending the available bandwidth to meet the requirements regarding the TAT. In turn, every additionally introduced data channel requires two further pins on the chip-level, whose maximal number is strictly limited by, among others, the pad-frame design. Thus, this number has to be chosen per design individually since it forms a trade-off between data bandwidth and costs.

Codeword-Based Compression
One important criterion of compression techniques concerns the completeness of the underlying compression algorithm. In this context, the completeness of the compression means that every arbitrary binary sequence can be processed. Consequently, none of the possible sequences, i.e., test patterns, will be rejected. By this, test coverage loss is prevented simply by design. As introduced in Chap. 4, SBIs are inserted into the compressed data to ensure the completeness of such a codeword-based technique with even small-sized embedded dictionaries. This embedded dictionary holds a particular number of codewords, which depends on the actual dictionary size. The dictionary implements a mapping function Ψ, whereby every single (short) binary codeword cdw_i is projected to a (long) binary dataword udw_i. Analogously to other techniques, the original test data have to be post-processed by a retargeting tool, which emits the compressed data, i.e., a sequence of codewords $cdw_1 ... cdw_n$. This sequence can be restored codeword-wise by Ψ to the original test data without any loss of information on-chip, which requires an implementation of Ψ as a part of the codeword-based decompressor in hardware. Individual codewords can be configured dynamically prior to the data transfer to further improve the compression effectiveness, which is realized by the DDU.

7.2 Hybrid Compression Architecture

Embedded compression techniques reduce the TDV significantly, which allows coping even with the large test sets of highly complex IC designs. However, a certain share of the overall test patterns is sometimes rejected by the compression infrastructure leading to a shortcoming that jeopardizes the zero defect policy for safety-critical applications. The proposed hybrid compression architecture tackles this drawback by combining a state-of-the-art embedded compression technique with a codeword-based approach, which is meant to be applied for the RTPs exclusively. By this, the significant overhead introduced by the RTPs can be effectively reduced while, at the same, keeping the fault coverage high.

7.2.1 *Motivation*

The basic idea of the proposed approach is about introducing a codeword-based decompressor instead of a simple bypass module as generally applied to address RTPs. The proposed scheme is shown in Fig. 7.2 and focuses on the incoming-data of the RTPs, which have to be retargeted once prior to the test application. The retargeted patterns, i.e., a sequence of codewords, are transferred bit-wise to the circuit. When a codeword is completed, the decompressor expands it to the (original) dataword and temporarily stores it until enough data are available to feed every input of the scan chains simultaneously. This is a significant improvement over the regular bypass, since it can feed the test data into the scan chains in **parallel** and not serially as the regular bypass does.

Three different modules are required to implement the proposed hybrid architecture, as described in the following:

1. The Codeword-based Decompressor implements the embedded dictionary and provides an interface to the mapping function Ψ,
2. the Hybrid Controller controls the operations of the codeword-based components, which includes the newly introduced control signal hybrid_en that configures the external datapath, and ensures that the hybrid compression components behave transparently if not activated, and
3. the Interface Module implements the junction between the newly introduced codeword-based decompressor and the inputs of the scan chains.

7.2.2 *Codeword-Based Decompressor*

This module implements the codeword-based technique utilizing an embedded dictionary to decompress the RTPs on-chip without any loss of information. Reconsider that these RTPs have been compressed once by the retargeting procedure prior to the test, i.e., the compressed data consists of a sequence of codewords $cdw_1, \ldots, cdw_n$. A method to determine efficient codewords for given test data has been previously presented in Chap. 5.

One integral component is the dictionary, which holds the codewords cdw_i in conjunction with the associated (uncompressed) dataword udw_i. A maximal codeword length of three bits is assumed in this paper since this bound has been identified as a good trade-off between compression effectiveness and hardware costs. Consequently, the embedded dictionary holds $\sum_{i=1}^{3} 2^i = 14$ entries, which

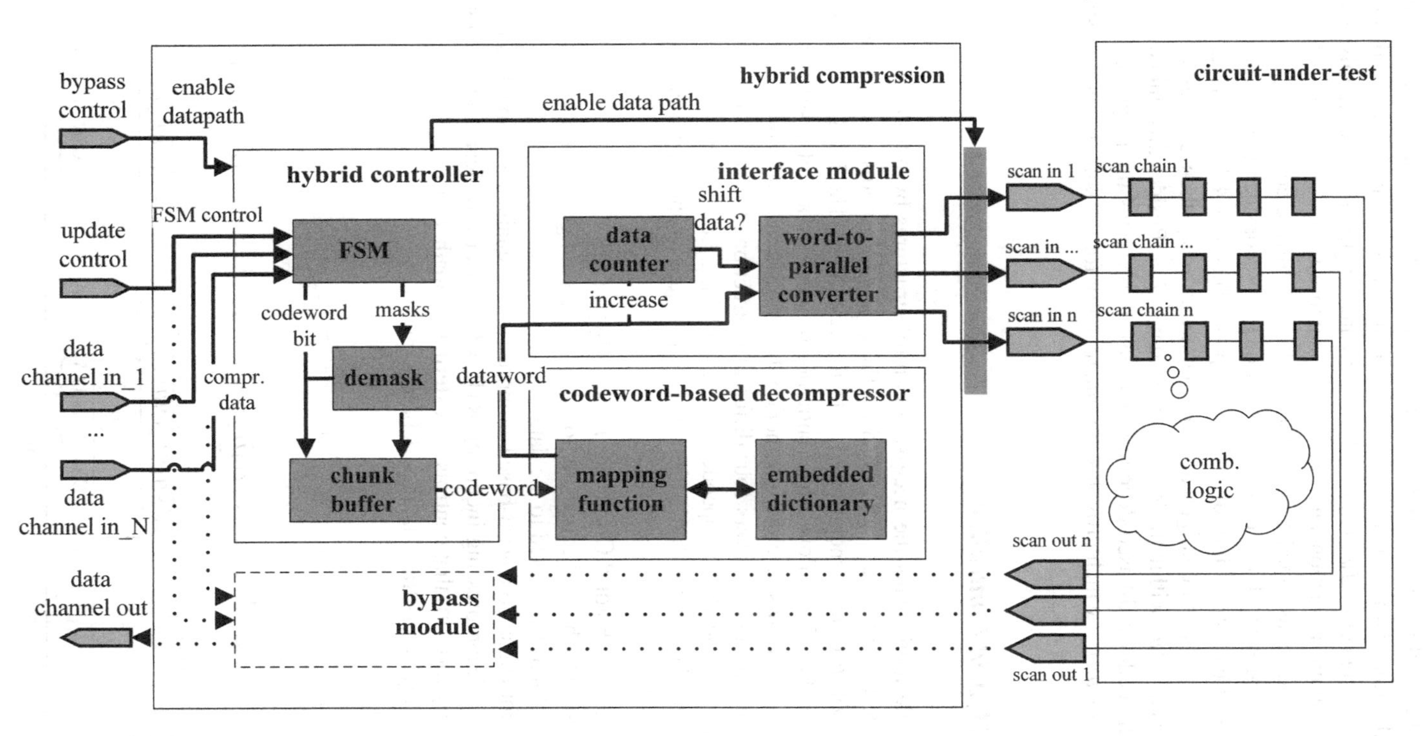

Fig. 7.2 Hybrid compression architecture

can be dynamically configured. Every entry consists of one binary-encoded, unique codeword cdw_i (with a length of 1 to 3 bits), and an associated (binary) dataword udw_i. This dataword udw_i tends to have a greater length $|udw_i|$ since this enables the data volume reduction and, hence, the compression. A maximum dataword length of 8 bits is assumed to keep the resulting hardware costs low, particularly, concerning the size of the embedded dictionary. Reconsider the basic principle of the cost function as introduced in Chap. 4, Theorem 4.10 (on page 64): If the codeword cdw_i is configured to the dataword udw_i, then each and every of occurences of udw_i within the RTP is replaced by cdw_i. Besides this embedded dictionary, the codeword-based decompressor also implements the mapping function $\Psi(cdw_i) \rightarrow udw_i$, which provides the functionality for the latter decompression.

7.2.3 *Hybrid Controller*

This module implements the necessary control structures by introducing a FSM, whose state transitions are controlled by the external compression control signal (update_control) and synchronized with the test clock. The FSM design, which is shown in Fig. 7.3, allows differentiating between data and instruction (*inst*) branch, which allows a clear distinction between data- and control-path. The current implementation covers four different instructions and, consequently, these instructions are encoded by opcodes holding two bits. However, this length can be easily extended to support further instructions if required. The supported instruction are as follows:

"00"' inactive; Hybrid Controller remains bypassed,
"01" preload; the embedded dictionary can be configured with specific codewords, which are adapted to the data,
"10" compress; codeword-based compression is enabled (disabled) and prepared for decompressing incoming RTP, which is also indicated by setting the signal (hybrid_en) that controls the external datapath, and
"11" debug; offers trace data for debug purposes.

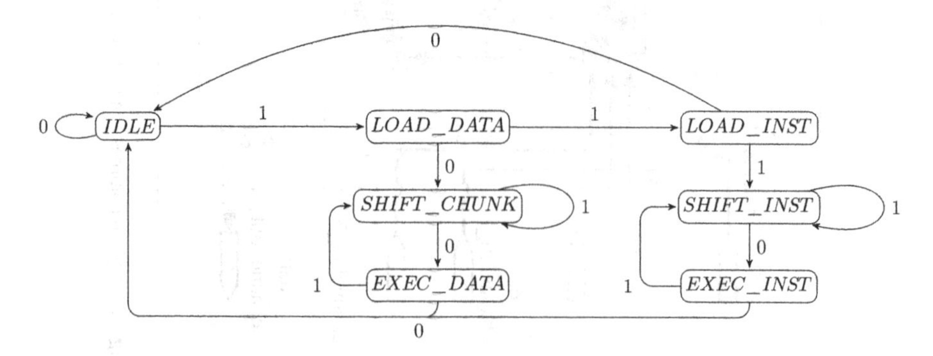

Fig. 7.3 Simplified FSM of hybrid controller: update_control @ edges

One crucial criterion of the overall design of Hybrid Controller is about the datapath of Hybrid Controller, which remains completely isolated until the bypass control signal is set. This principle allows avoiding any unintended interference when the regular compression infrastructure decompresses the unrejected test patterns.

Loading Instructions

The FSM remains in the idle state as long as the bypass and compression control signal are both set. To execute an instruction, the specific opcode has to be bit-wise transferred to Hybrid Controller via the data input channel while recursively traversing the state shift_inst twice. This state can be reached by passing both load_data as well as load_inst states first. After the transfer is completed, the transferred opcode is decoded and, finally, executed within the state exec_inst. This instruction cycle ends up in the idle state or at shifting further instruction. If the instruction compress has been executed, the newly introduced hybrid_en signal is set. This signal controls the external datapath with respect to the inputs of the scan chains.

Loading Data

The compressed RTP consists of a sequence of codewords $cdw_1, \ldots, cdw_n$. Each and every of the n codewords have to be transferred in a bit-wise fashion. Analogously to the instruction load, the bit-wise transfer for the i-th codeword cdw_i is executed while recursively traversing the state shift_chunk. A bit of a codeword (chunk) is transferred with each iteration and stored in a dedicated chunk buffer. The state is changed to exec_data as soon as all bits of the current codeword cdw_i are completely transferred. The actual decompression is executed within this state by orchestrating the mapping function Ψ. This state separation allows identifying the end of a single codeword uniquely. The next reached state is then either shift_chunk to start a further codeword transmission or idle to finish.

7.2.4 Interface Module

This module realizes the junction between the introduced Codeword-based Decompressor and the T inputs of the scan chains. In this scenario, the decompressor acts as the data source by decompressing newly received codewords to the associated datawords using Ψ and the inputs of the scan chains act as the data sink. This module is especially important to feed the scan data into the scan chains in parallel differently to the regular bypass, which works serially. As stated, the codewords of the embedded dictionary can be configured to arbitrary datawords (within the assumed boundary). Consequently, the length of the individual dataword is not necessarily the same, which ensures a high flexibility and, thus, achieves high compression effectiveness. Due to this fact, a mechanism is required that keeps track of the actual number of available data bits. If enough data bits are available, the

parallel shift operation to all scan chains is performed or, otherwise, more data bits are received until the required amount of data is available.
Three different components are required to implement Interface Module as follows:

1. A data counter keeping track of the currently decompressed datawords, i.e., the number of available bits at this point of time,
2. a capture register storing the currently decompressed bits, and
3. an update register holding the bits that are shifted into the scan chains.

After all chunks of a codeword cdw_i have been received, cdw_i is decompressed and stored in capture register. The counter is then increased by the length of the newly decompressed dataword meaning $|udw_i|$. If the counter exceeds the required value of T, the parallel shift operation is performed by transferring the T least recently received bits from the capture register to the update register, which is then used as the data source. The counter is then decreased by T.

7.2.5 Exemplary Application

The remainder of this section describes the exemplary application of the hybrid compression technique. For the latter example, it is assumed that the components of the compression architecture as well as of the hybrid compression have been both implemented in the circuit and the test pattern set has been generated. Furthermore, it is assumed that the RTPs have been identified and retargeted by the utilized retargeting framework.
The utilization of the hybrid compression is as follows (consider t as the point of time in sense of passed cycles):

1. $t = 0$: The external bypass control signal is set to “1” and remains in this state, which activates the datapath of Hybrid Controller.
2. $t = 1..3$: The external update control signal is set to “1” for three clock cycles to, finally, reach the state shift_inst.
3. $t = 4..6$: The opcode “10” is shifted into Hybrid Controller by controlling the data input channel to “1” ($t = 4$) and to “0” ($t = 5$). The transferred instruction is decoded and executed in the state exec_inst ($t = 6$). This implies that the hybrid_en signal is set, which activates the datapath between Interface Module and the input of the scan chains and, thus, deactivates the classic bypass architecture.
4. $t = 7..9$: The update control signal is driven such that the state shift_chunk is reached.
5. $t = 10..11\text{+x}$: Transfer of the first codeword, which requires at least one but up to three cycles depending on the number of chunks. The data bits are transferred via the data input channel.

6. t = 12+x..13+x: The first codeword cdw_i is completed, which is indicated by the state transition to exec_dates (induced by the control signal change "1" to "0"). The decompression is performed by invoking $\Psi(cdw_i)$ to restore the dataword udw_i with respect to the embedded dictionary. Subsequently, udw_i is directly stored in the capture register and the data counter is increased by $|udw_i|$.
7. t = 14+x: If a further codeword cdw_{i+1} has to be transferred, the state is changed to shift_chunk and it is continued with Step 5). Otherwise, the state is changed to idle. The hybrid compression can be deactivated again by repeating Step 2) as well as Step 3); however, while loading the opcode "00." This instruction deactivates the datapath between Interface Module and the inputs of the scan chains. The complete datapath of Hybrid Controller is isolated by applying Step 1) in reverse.
   In parallel, the value of the counter of Interface Module is evaluated: If counter $\geq N$ is valid, and the least recently received T bits are moved from the capture register to the update register such that a parallel shift operation is performed (t = 15+x). Furthermore, the counter is decreased by T.

The individual codewords of the embedded dictionary can be configured by first executing the preload instruction [cf. Step 2), 3)] and, second, the datawords have to be transferred in a bit-wise fashion—analogously to Step 4) to 6). This configuration phase is not discussed in more detail to keep the exemplary application short and, more importantly, since the basic principles of the configuration have been demonstrated in Sect. 6.1.2 on page 96.

7.3 Extended Hybrid Compression

An important aspect concerns the estimated hardware overhead, which is introduced if an arbitrary circuit design implements the hybrid compression technique. Furthermore, it can be observed that industrial designs often employ multiple (data) channels for the high volume manufacturing test. The remainder of this section discusses both of these aspects, first to formulate a cost metric and, second, to present an extension of the previously proposed hybrid compression scheme to take advantage of an existing multichannel topology.

7.3.1 Hardware Cost Metric

The hardware costs of Codeword-based Decompressor and Hybrid Controller are both completely independent from the design size but are mainly determined by the parameters of the embedded dictionary. These parameters are the MCL defining the upper bound for unique binary encodings of the codewords and the MDL determining the word size of the single dictionary entry. As stated, MCL is set to 3 resulting in maximum 14 entries. More precisely, the dictionary holds two entries with a length equal to 1, four entries with a length equal to 2, eight entries with a length equal to 3, etc. Generally spoken, **(a)** $\sum_{i=1}^{MCL}(2^i \cdot i)$ bits are required to store all codewords. Furthermore, the associated dataword (with a maximal length of MDL bits) for all possible codewords has to be considered with **(b)** $\sum_{i=1}^{MCL}(2^i \cdot MDL)$ bits.

In contrast to this, the costs of Hybrid Controller scale with the number of considered inputs of scan chains (T). The size of the update register has to be equal to **(c)**T since this is the exact number of bits that are required to perform the parallel shift operation. The received and decompressed bits are stored in the capture register, which has to hold a size of **(d)** $T + MDL - 1$ bits. This covers the scenario that CR is filled with $T - 1$ data bits (the critical amount for performing the parallel shift operation is not yet reached) and a further dataword of MDL. Additionally, Interface Module contains an (unsigned) data counter of **(e)** $\lceil \log_2(T + MDL - 1) \rceil$. Some extra registers are required to hold the current state and the loaded instruction of Hybrid Controller and, finally, chunk-buffer holding **(f)** MCL bits. Accumulating (a) to (f) results in

$$\underbrace{\sum_{i=1}^{MCL}(2^i \cdot i)}_{(a)} + \underbrace{\sum_{i=1}^{MCL}(2^i \cdot MDL)}_{(b)} + \underbrace{T}_{(c)} + \underbrace{T + MDL - 1}_{(d)}$$

$$+ \underbrace{\lceil \log_2(T + MDL - 1) \rceil}_{(e)} + \underbrace{MCL}_{(f)}$$

$$= \underbrace{\sum_{i=1}^{MCL}[2^i \cdot (i + MDL)]}_{(a)+(b)} + \underbrace{\lceil \log_2(T + MDL - 1) \rceil}_{(e)}$$

$$+ \underbrace{2T + MDL + MCL - 1}_{(c)+(d)+(f)}$$

with MCL="3" and MDL="8"

$$2N + \lceil \log_2(N + 7) \rceil + 156$$

As shown in (2), the hardware costs scale super-linearly with the number of scan chains. In industrial-sized designs with a vast number of scan chains, the individual input is not accessible and multiple scan chains are tail-gated. This tail-gating massively reduces the number of scan chain inputs, which have to be directly driven. For the sake of simplicity, this technique has not been considered in this calculation.

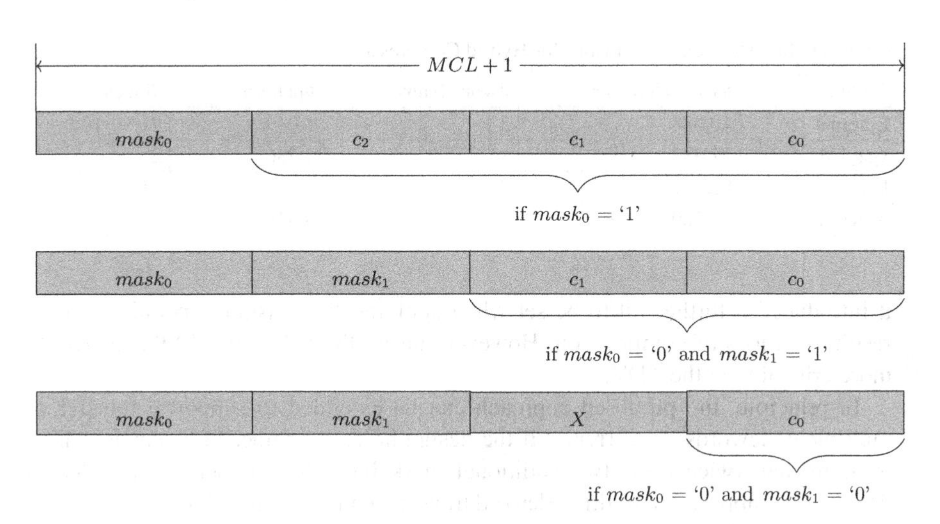

Fig. 7.4 Masking scheme of multichannel topology

7.3.2 Multichannel Topology

The transfer of all compressed datawords $cdw_0, ..., cdw_n$ is implemented in a sequential fashion as described in the exemplary application in Sect. 7.2.5. More precisely, each compressed dataword cdw_i contains up to MCL (here 3) chunks c_i with $i < MCL$, which are sequentially transferred while iterating over the state shift_chunk up to MCL-times. This number of required iterations can be statically reduced to 1 by utilizing existing multichannel structures, which typically exists in large designs [Jan+11]. To achieve this, $MCL + 1$ channels are orchestrated to transfer all chunks of a single compressed dataword in parallel while occupying one additional bit implementing a mask (Fig. 7.4).

Assume that four channels $[3 : 0]$ are available and the current compressed dataword consists of 3 bits (chunks). In this scenario, the chunks of the codeword are given by $[2 : 0]$. The information about the relevant bits has to be encoded in the transfer since this is important for the interpretation of the data. This is achieved by the introduction of a masking scheme, which reflects the actual number of relevant channels for the current compressed dataword under transfer. The other bits can be discarded and, hence, they are handled as X-values.

Considering the number of available channels as $MCL + 1$-bit-wide bus $data[MCL : 0]$, the most-significant bit $data[MCL]$ indicates whether all the remaining $data[MCL - 1 : 0]$ are used to transfer a codeword of length MCL. If $data[MCL]$ is set, the codeword is transferred by $data[MCL - 1 : 0]$. Otherwise, the $data[MCL - 1]$ acts as this mask for the remaining $data[MCL - 2 : 0]$, which indicates that a codeword of length $MCL - 1$ is transferred.

It is assumed that at least $data[0]$ holds a chunk of a codeword with length 1. This scheme allows a further reduction regarding the test application time, though,

Table 7.1 Industrial circuit statistics for Hybrid Compression

Circuit	#scan. elements	#scan chains	#pattern	#blocks
Ethernet	10,038	5	1,049	2
vga_lcd	12,983	14	1,286	2
leon3	82,251	81	6,332	11
Netcard	96,569	97	9,939	12

it introduced a further bit to be set when applying the masking operation, which results in a loss of compression. However, during the LPCT the TAT is generally more critical than the TDV.

In principle, the proposed approach can be extended to support a transfer of multiple codewords concurrently if the design holds a number of channels, which is more than twice (plus two additional mask bits) the number of the selected MCL. The support of a multi-codeword transfer requires some additional hardware components, which have to be introduced into the design. For instance, a separate register is required to store the chunks of the additional compressed dataword.

7.4 Experimental Setup

This section describes the experimental evaluation of the proposed technique, which has been done by using different industrial representative OpenCores as well as Gaisler Research Benchmark circuits from the IWLS 2005 benchmark collection [Alb05].

The underlying scan capabilities and the embedded test compression have been inserted by a commercial tool. The characteristics of the resulting circuits are shown in Table 7.1 as follows: The benchmark circuit name (column circuit), the overall number of scannable sequential elements included in the design (column #scan. elements), the number of introduced scan chains[1] (column #scan chains), the overall number of test patterns, which have been generated by the commercial tool (column #pattern), and the number of required data blocks per RTP (column #blocks). In the case of the leon3 and netcard circuits, a multichannel topology exists. Thus, the newly proposed multichannel architecture is evaluated based on these two industrial representative candidates.

As indicated earlier, the number of RTPs might be significant [Dho+16]. However, it depends on the test application and also may differ between different synthesis runs. In order to have reproducible results, we, therefore, consider in the following a large share of the test patterns for the experimental evaluation. The hardware costs of the proposed hybrid compression architecture are within the same

[1]Note that a maximal scan chain length of 1,024 elements is considered.

order as the regular embedded compression hardware, which has been validated by a commercial synthesis tool.

The proposed codeword-based compression module is solely implemented in Verilog and the developed module is then instantiated within Hybrid Controller, whereby various Verilog Parameters [01] allow an easy adoption to further benchmarks. The test patterns, which are generated by a commercial tool, are then processed by the retargeting framework of Chap. 5, which has been slightly adopted. Among others, it is combined with a partitioning approach of Chap. 6 to keep the run-time manageable while further improving the effectiveness. Thus, data blocks with a size of 8k are considered for the retargeting. The retargeting is executed on a Fedora 30 (with gcc 9.2.1) system holding an Intel Xenon E3-1240v2 3.4 GHz processor with 32 GB system memory. Finally, a commercial tool has been used to validate the (simulated) transfer and decompression on-chip by the newly introduced hybrid compression module.

7.5 Experimental Results

The detailed results of the conducted experiments are presented in Table 7.2 as well as in Fig. 7.5. More precisely, the benchmark circuit name and the minimal, average and maximal values for the retargeting run-time per pattern in seconds and the TDV as well as TAT reduction in percentage. Note that the configuration of the embedded dictionary is applied once prior to the data block transfer. Both the configuration time and the configuration bits are included in the results. Furthermore, the proposed hybrid compression module is complete by construction, i.e., any arbitrary RTP can be transferred and, thus, no test coverage loss is introduced. In the case of the *netcard* benchmark, at first, the experiments clearly show that the run-time of the retargeting engine, while considering different design sizes, is stable **per** block. Since the retargeting has just applied once after the test pattern generation process, the run-time is manageable. Furthermore, the retargeting invokes currently only one single thread and the individual blocks are objectives to be parallelized. Second, the achieved compression ratio is stable over the conducted experiments as well, which is indicated by a variance standard deviation of 6.4% (netcard). In contrast to this, the deviation between the minimum and the maximum of achieved

Table 7.2 Hybrid compression: TDV &TAT reduction of industrial circuits

	Retargeting run-time [s]			TDV red. [%]			TAT red. [%]		
Circuit	min.	avg.	max.	min.	avg.	max.	min.	avg.	max.
Ethernet	13.2	26.4	47.2	31.1	56.4	79.3	34.6	55.5	72.8
vga_lcd	13.6	37.6	42.1	30.6	40.7	56.1	30.5	41.5	47.9
leon3	64.3	293.7	513.2	31.7	48.7	79.2	27.2	48.8	69.4
Netcard	104.4	126.0	230.4	28.5	38.1	67.4	29.6	45.7	65.7

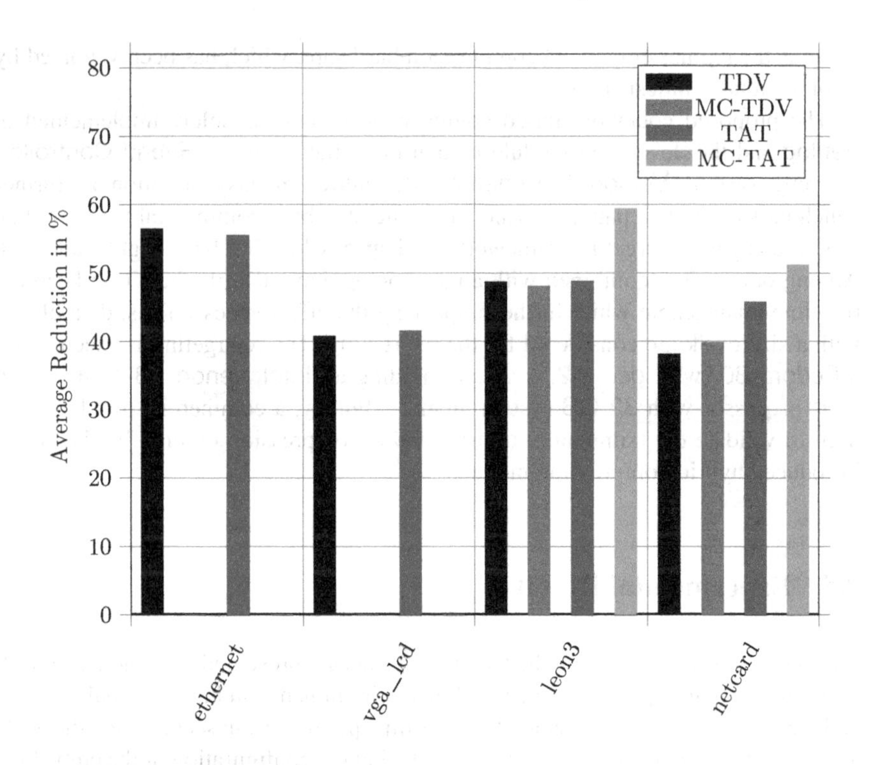

Fig. 7.5 TDV & TAT of hybrid (multichannel) compression

test time reduction is higher, which is because this depends on the distribution of codewords with a length of 1, 2, or 3, respectively. Thus, the test time reduction is not directly connected with the compression ratio. The latter is determined mainly by the associated datawords, to which the introduced codewords are pointing to. When processing the largest netcard circuit with the highest number of patterns ($N = 9939$), the resulting test data volume can be significantly compressed by 38.1% on average in conjunction with a test time reduction of 45.7% on average.

The results of the conducted experiments with multichannel topologies are presented in Table 7.3, which gives the number of available channels on the chip-level (column #channels) and the minimal, average, and maximal reduction ratios in percentage for the TDV as well as the TAT. The ratio of scan chains divided to introduced channels has a substantial impact on the share of RTPs. Reconsider that specific applications like the LPCT holds a very high ratio. Further note that the approach requires more than one channel and, hence, this approach has been evaluated based on the extra-large leon3 as well as netcard benchmark circuits. The results clearly demonstrate that the TAT can be further reduced by up to 72.9% while still achieving a compression ratio up to 67.4%.

Table 7.3 Hybrid Multichannel Compression: TDV &TAT reduction of industrial circuits

Circuit	#channels	Multichannel TDV red. [%]			Multichannel TAT red. [%]		
		min.	avg.	max.	min.	avg.	max.
Ethernet	1	n/a	n/a	n/a	n/a	n/a	n/a
vga_lcd	1	n/a	n/a	n/a	n/a	n/a	n/a
leon3	4	14.8	48.1	77,9	38.4	59.4	84.8
Netcard	4	20.1	39.8	48.9	34.6	51.2	72.9

7.6 Summary

This chapter proposed a hybrid architecture, which seamlessly combines a state-of-the-art embedded compression technique with a lightweight codeword-based compression scheme to enable the processing of rejected test patterns. So far, these RTPs had an adverse impact on the overall compression effectiveness and, hence, an untapped potential of reducing the test costs remained. The proposed architecture introduces only a negligible hardware overhead in the sense of register count to the design and the computational effort for the preprocessing step, i.e., the retargeting, is manageable.

Different experiments on industrial representative designs—with up to approximately 100k flip-flops—demonstrate the effectiveness of the proposed hybrid architecture. A significant TDV reduction of up to 67.4% was achieved and, most important, without any loss in test coverage at all. Furthermore, the TAT was reduced even stronger by up to 64.7%. In addition to this, the proposed hybrid architecture allows utilizing existing multichannel structures. By this, it was possible to further decrease the TAT by up to 72.9% with up to 48.9% of TDV reduction. In particular, the massive decrease of the TAT saves test costs. The proposed hybrid compression architecture forms the new baseline in the sense of test compression for LPCT applications.

Chapter 8
Enhanced Reliability Using Formal Techniques

Several breakthroughs in the field of the design, fabrication, and test of integrated circuits allowed for the implementation of highly complex ICs. These ICs fulfill several mission- or even safety-critical tasks at once while following a highly complex functional behavior.

The enormous complexity and the intensive environmental interaction, particularly, in the case of long-term autonomous systems, lead to new challenging requirements with respect to the reliability of IC designs. Moreover, shrinking feature sizes as well as different environmental influences such as high-energetic radiation, electrical noise, particle strikes, etc. frequently cause an unintended behavior of an IC, which can lead to disastrous consequences in the worst case. The sequential components of the IC, meaning flip-flops, are characteristically vulnerable to the so-called transient faults. Such a fault appears in the form of a toggled bit for a short period of time. If this fault is neither logically, electrically nor temporarily masked, the output signals of the system are invalidated and, hence, the functional correctness is harmed. In particular, when designing ICs for safety-critical applications, the contained FFs have to be explicitly protected to avoid such an incorrect functional behavior. Powerful mechanisms have to be developed to protect the IC designs and, hence, to address this highly relevant field of research. These mechanisms are introduced into the IC design and allow detecting or reacting on a recently occurred transient fault.

In this context, the robustness of a given circuit is an important metric, which can be derived from the number of non-robust FFs that are vulnerable to transient faults. The number of vulnerable (non-robust) FFs has to be decreased to increase the robustness of a sequential circuit. To this end, the corresponding FFs are *hardened* by extending the investigated circuit such that the respective values are recomputed and, in the case of faults, faulty signals can be replaced. These recomputations are usually conducted either by additionally employing redundant hardware or redundant time. More precisely, existing methods which are currently

S. Huhn, R. Drechsler, *Design for Testability, Debug and Reliability*,
https://doi.org/10.1007/978-3-030-69209-4_8

applied to increase the robustness of a given sequential circuit can roughly be categorized into the following three schemes [KCR06]:

- **Space-based** approaches, which embed additional logic blocks to generate certain redundancy to enhance their robustness. Approaches such as Triple Modular Redundancy [BW89, SB89] or Error-Correction Code [Luo89, Ham50, SRM14] constitute representatives of this scheme.
- **Timing-based** approaches, which influence the timing behavior of the considered circuit to guarantee correct output values at the FFs. Representative candidates of this scheme have been proposed in [Ern+03, Bla+08].
- **Application-specific** approaches, which only consider dedicated parts of a circuit for which a robust solution is explicitly derived. Examples of this scheme include, for instance, dedicated fault-tolerance control flows for microprocessors [FFM08]. Other application-specific approaches exploit invariants automatically to ensure correctness like hardware assertions [HSV13]. Those assertions are used to uncover violations of the specific functional behavior during verification but do not focus on any dedicated fault model considering transient faults and, hence, no compact realization is given a priori.

However, all these schemes have significant shortcomings: Space-based approaches introduce huge additional logic into the circuit—mostly due to a naive multiplication of sequential elements and, hence, a redundancy which is not required for the functional behavior at all. Timing-based approaches suffer from the fact that they potentially increase the latency, which heavily constrains the design of the circuit. Application-specific approaches are limited to dedicated parts of a circuit, for instance, the considered microprocessor, implying that are not applicable for arbitrary sequential circuits.

This chapter proposes a novel technique to address these shortcomings, which does not merely recompute FFs values of the original design, but determines application-specific knowledge of their behavior. More precisely, the proposed methodology determines the relations of FFs and stores the conditions under which FFs assume the same logic value to partially harden the FFs. This knowledge allows employing a dedicated logic block which compares all corresponding FF values and, hence, can detect if one of them inherits a fault. Since multiple blocks are introduced—as described later in more detail—the proposed approach even enables the detection of multiple transient faults under certain conditions. The proposed application-specific knowledge is derived out of typical signal values, which occurs in the intended functional operation. Conventional techniques as described above cover the entire state space exhaustively, i.e., those techniques cover even a high fraction of irrelevant state space. By this, redundancies within the IC are exploited. Even if, different optimization techniques are applied in a typical IC design flow to eliminate redundant components, often a significant amount of functional equivalence remains and can be exploited for the purposes considered here.

As an example, consider a Full Adder with three inputs (a, b, c) and a Half Adder with two inputs (a, b). Both of adders have two outputs representing the sum and the carry-out bit. Although these circuits are not entirely functional equivalent (both indeed realize slightly different functions), their behavior is identical whenever the carry input c is set to "0"—the circuits are partially functional equivalent and generate the same value in specific states. Having the information on what signals assume the same value in which state can be used to compare those signals with each other and, by this, to strengthen the robustness of sequential circuits. Note that a more detailed description of the proposed idea is later given in Sect. 8.1. This yields significant improvements compared to the other solutions reviewed above: Exploiting this information on partial redundancies allows increasing the robustness with only a moderate hardware overhead in terms of gate count compared to the space-based approaches. At the same time, the timing behavior is only affected negligibly, particularly, in contrast to the timing-based approaches. The proposed flow is automatically applicable to any sequential circuit and, hence, not limited to dedicated circuits as the existing application-specific solutions. The developed methodology needs to determine the needed information, i.e., the relations of FFs and the conditions under which they assume the same logic value. Since this is a computationally hard task, we propose to address this with a dedicated orchestration of formal methods such as BMC [Bie+99b], powerful solvers for the SAT problem [ES04], SAT-based ATPG [ED10], and compact data structures involving BDD [Bry86], which is capable of coping with this complexity Finally, this approach is flexible in the sense that the designer can easily configure the trade-off between the hardware overhead and the desired enhancement in robustness.

The structure of this chapter is organized as follows: Sect. 8.1 introduces the proposed methodology in detail while distinguishing against related works and, furthermore, draws the general idea by introducing a motivating example. In Sect. 8.2, the core components are successively introduced, which are the Partition Enumerator, State Collector and, furthermore, the resulting FDM. The experimental setup is described in Sect. 8.3, which includes the simulation-based robustness checker. The experimental evaluation of this newly proposed knowledge-based transient fault detection methodology is presented in Sect. 8.4. Finally, Sect. 8.5 summarizes the paper and gives an outlook on future work.

8.1 Motivation

The idea of the proposed methodology is inspired by the fact that today's circuits usually contain a vast number of FFs, which can store at least a single bit, i.e., "0" or "1." If a single FF is affected by a transient fault, this bit is toggled. Existing approaches insert redundant logic into the design, e.g., to recompute the correct value, which causes a significant hardware overhead. At the same time, the value of an observed single FF is often equal to the value of many other FFs. It is possible to determine the relation between them, i.e., the states in which certain FFs assume the same value since the behavior of the circuit is known. Instead of introducing redundancy for recomputations, it is proposed to compare the value of a FF to the

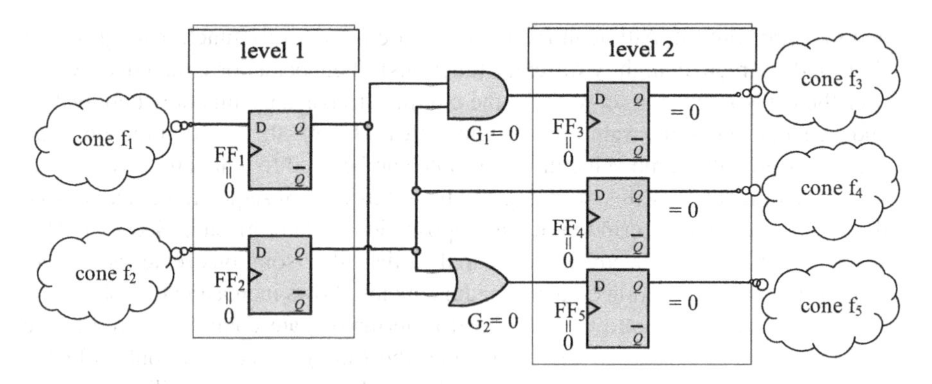

Fig. 8.1 A non-robust sequential circuit

values of other FFs from which it is known that, for the respectively considered state, they are supposed to generate the same (output) value. A formalism is required to realize this idea that states whether a partition of non-robust FFs assumes the same value for given reachable states. This is formally described as an *Equivalence Property* (EP), as stated in Theorem 8.1.

Definition 8.1 Let $P_j \subseteq N$ be a partition of at least two non-robust FFs and $\widehat{S} \subseteq S^*$ be the set of reachable states, whereby S^* represents the complete state space and $\widehat{S}$ reflects the reachable state space of a given sequential circuit. Then, an Equivalence Property (EP) is defined by

$$\mathsf{EP}(\widehat{S}, P_j) := \left\{ f_1, \ldots, f_l \in P_j \;\middle|\; \begin{array}{l} \text{all FFs } f_1, \ldots, f_l \text{ outputs the same value} \\ \text{under the same state } s \in \widehat{S} \end{array} \right\}$$

and evaluates to true if all combinations of FFs $f_n, f_m \in P_j$ assume the same output value in all of these states $\widehat{S} \subset S^*$.

Example 8.2 Consider the circuit shown in Fig. 8.1, which is composed of five FFs distributed in two hierarchical circuit levels 1 and 2. If both FF_1 and FF_2 (level 1) are set to "0," then FF_3, FF_4, and FF_5 (level 2) are assumed to have the same output value "0" after a single clock cycle. This scenario is represented by an $\mathsf{EP}(\widehat{S}, P_j)$ with the partition $P_j = \{FF_3, FF_4, FF_5\}$ and the state $s_j \in \widehat{S}$ being defined by $FF_1 = 0$ and $FF_2 = 0$, i.e., $\mathsf{EP}(\widehat{S}, P_j) = 1$ holds.

An elaborated orchestration of the EP concept in combination with the sketched idea above allows to enhance the robustness of sequential circuits as follows:

1. Determine the set $N \subseteq \mathsf{SE}$ of non-robust FFs of the given sequential circuit (cf. Theorem 2.2 on page 12). The assessment of robustness as reviewed in Sect. 2.6.1 on page 30 can be utilized.

2. Consider the *set of non-robust FFs* (N) and determine the level-wise subsets $N_i \cup N_{i+1} \cup \cdots \cup N_L = N$ with $1 \leq i \leq L$ according to their hierarchical circuit levels (L being the total number of hierarchical levels in a rank-ordered circuit [Mic03, p. 45]). Furthermore, assume that each FF has exactly one hierarchical level: $N_i \cap N_j = \emptyset \; \forall i \neq j$. The above-described clustering allows executing all FFs' comparisons within one single time frame. Thus, this clustering is crucial to reduce the complexity of the calculation itself and, especially, the costs of the robustness improvement. Consequently, there is no need to hold specific FF values over different time frames, e.g., by introducing further—potentially vulnerable—state elements while massively increasing the computational effort as well as hardware scale. Thus, the level-wise sets of non-robust N_i are exclusively used in the remainder.
3. For each level $1 \leq i \leq L$ and for all subsets of non-robust FFs $N_i \subseteq N$, determine suitable partitions $P_j \in \mathcal{P}(N_i)$ and a set of reachable states $\widehat{S} \subseteq S^*$ such that all FFs in P_j are supposed to generate the same value. This means that a set of partitions is determined (with their individual $\widehat{S}$), for which $\mathsf{EP}(\widehat{S}, P_j) = 1$ holds.
4. Use the knowledge from the obtained EPs and synthesize a FDM. To this end, realize the following logic blocks:
   - Activator: Generates the *Activator signal* ($\mathcal{A}$)—supposed to trigger the FDM—stating whether ($\mathcal{A} = 1$) or not ($\mathcal{A} = 0$) the FFs in P_j are supposed to generate the same value under the current state $s \in \widehat{S}$. This signal is directly calculated by the current state (single time frame) of the FFs within the fan-in cone. More precisely, it is not required to consider previous values, which is solely enabled by the hierarchical sort as described above in Step 2).
   - Equality Comparator: Generates an *Equality Comparator signal* ($\mathcal{E}$) stating whether ($\mathcal{E} = 1$) or not ($\mathcal{E} = 0$) all FFs of a given partition P_j, which are considered for hardening, assume the same output value.
   - Detector: Generates a *Fault signal* ($\mathcal{F}$) reporting the detection of a fault. A fault is detected if not all FFs in a partition P_j assume the same output value ($\mathcal{E} = 0$), although they are supposed to do that for the current state ($\mathcal{A} = 1$) meaning $\mathcal{F} = \neg\mathcal{E} \bullet \mathcal{A}$.

This proposed FDM detects transient faults occurring in FFs of the considered circuit. The introduced $\mathcal{F}$ is driven if a fault is detected. This enables the implementation of precautions against faulty behavior at the POs, for instance, by resetting the circuit or masking the affected POs. Since the signal $\mathcal{E}$ is directly driven by the output of the hardened FFs, the proposed approach allows detecting silent data corruption of these FFs (with one clock cycle of delay), which are not yet visible at a PO. Overall, this leads to enhanced robustness. The ratio of the enhancement can thereby be controlled, for instance, by adjusting the amount of knowledge collected through the EPs.

8.2 Application-Specific Knowledge

This section describes the methodology to determine application-specific knowledge of an arbitrary sequential circuit. This knowledge is then used to generate a circuit-specific FDM, which allows identifying recently occurred transient faults.

The determination of—as much as possible—application-specific knowledge in terms of EPs is one essential step of the proposed methodology. Ensuring the completeness would require that all possible partitions $P_j \in \mathcal{P}(N_i)$ of all non-robust FFs in the same hierarchical circuit level are considered. Obviously, this leads to an exponential complexity, which is not feasible for practical applications. Moreover, most of the partitions P_j are likely to be not suitable for an EP anyway since no state s_j may exist for them so that all assume the same value.

A mechanism is introduced, which aims at determining *good* partitions effectively. Mainly, a SAT-based ATPG model [ED10] as introduced in Sect. 3.4 is adopted to compute the criteria of quality for an investigated partition. In addition to that, formal methods are heavily exploited such as BMC [Bie+99b], powerful solvers for the SAT-problem [ES04], and compact data structures involving BDDs [Bry86].

This leads to a methodology composing the components as follows:

1. A Partition Enumerator selects suitable partitions $P_j \in \mathcal{P}(N_i)$, which have not been considered before.
2. A State Collector determines the states $\widehat{S}$ under which all FFs in the selected partition P_j assume the same value and, hence, determines all $\mathsf{EP}(\widehat{S}, P_j)$ evaluating to true.
3. An FDM Synthesizer takes the obtained knowledge, realizes the FDM, and embeds the resulting logic into the original circuit.

The remainder of this section describes each step in detail and illustrates it by means of the circuit given in Fig. 8.1.

8.2.1 Partition Enumerator Framework

The Partition Enumerator is supposed to select partitions of FFs, which are likely to generate the same value in as many as possible (reachable) states. All non-robust FFs (given in N) are initially distinguished according to their hierarchical circuit level. The FFs of a circuit level N_L are then divided into sets of partitions, which are separately considered. However, a method is needed which returns a proper set of partitions yielding *good* robustness results since it is not practically feasible to enumerate all possible partitions.

Two different methods are proposed in the following: The first method $\mathcal{P}_{\mathsf{RAND}}$ invokes a random-based computation and the second one $\mathcal{P}_{\mathsf{SAT}}$ exploits SAT-based techniques. However, both techniques differ strongly concerning their complexity

Algorithm 5 Partition enumeration procedure

Require: set of non-robust FFs: N_i and partition generator: $\mathcal{P}_{\mathsf{GEN}}$
1: Container $\mathcal{Z} = \emptyset$ {Data container for EPs}
2: Context ctx {Overall context}
3: **while** True **do**
4: Let $P_j = \mathcal{P}_{\mathsf{GEN}}(ctx, N_i)$ {PGEN calls PRAND _or_ PSAT}
5: **if** $P_j = \emptyset$ **then**
6: **return** $\mathcal{Z}$
7: **end if**
8: $\widehat{S} =$ StateCollector(P_j)
9: **if** $\widehat{S} \neq \emptyset$ **then**
10: $\mathcal{Z} = \mathcal{Z} \cup \mathsf{EP}(\widehat{S}, P_j)$
11: ctx.accept(P_j)
12: **else**
13: ctx.decline(P_j)
14: **end if**
15: **end while**

and quality. Generating a random set is considerably less time-consuming than solving a SAT instance. In contrast to this, the quality of the enumerated sets is much better following the SAT-based approach since the respectively guided search considers the functional behavior of the circuit.

A common algorithmic framework has been developed for both methods $\mathcal{P}_{\mathsf{RAND}}$ and $\mathcal{P}_{\mathsf{SAT}}$ as stated in Algorithm 5. The non-robust FFs N_i and a partition generator function $\mathcal{P}_{\mathsf{GEN}}$—sharing the algorithmic base of both methods—are provided as an input. The generator function additionally gets a context object representing the actual state and providing two functions: accept marks a partition to be accepted and decline marks that the State Collector could not identify any state satisfying EP, which is strictly required as stated in Theorem 8.1.

The approach evaluates new partitions as long as $\mathcal{P}_{\mathsf{GEN}}$ generates non-empty ones by invoking either the random-based $\mathcal{P}_{\mathsf{RAND}}$ or the newly proposed SAT-based $\mathcal{P}_{\mathsf{SAT}}$ technique based on the designer's choice. In contrast to the $\mathcal{P}_{\mathsf{RAND}}$ random technique, $\mathcal{P}_{\mathsf{SAT}}$ invokes the proposed metric to determine the quality (with respect to the stated criteria) of an arbitrary partition. This allows evaluating the quality of the individual partition to, primarily, select the best—again with respect to the stated criteria—one. In each iteration, the generated partition is passed to the State Collection (line 8) that computes the states $\widehat{S}$ in which all FFs in the currently considered partition P_j assume the same value (described in detail in Sect. 8.2.4). If at least one state $s \in \widehat{S}$ exists (lines 9 to 11), an EP is created using the states $\widehat{S}$ determined by the State Collector as well as the currently considered partition P_j. Then, the resulting EP is stored within a global data container $\mathcal{Z}$ (used later by the FDM synthesizer) and the partition P_j is accepted by calling accept. In contrast, if no state exists, i.e., $\widehat{S} = \emptyset$ the partition is declined by calling decline.

8.2.2 Randomized Partition Search

The first method $\mathcal{P}_{\text{RAND}}$ to obtain partitions selects some FFs of a given set randomly. Generating those partitions is very fast but does not consider any logical context of the FFs. However, this naive approach of randomly selecting partitions achieves already strong improvements in terms of robustness while keeping the hardware overhead low.

Technically, a set of non-robust FFs and a pre-defined upper bound p_s are provided as an input. The resulting size of the generated partition is less than the pre-defined bound. The algorithm randomly picks an uncovered FF and adds it to the partition. Finally, the partition is checked whether there are some states to satisfy EP. If there are such states, the partition is taken for hardening. Otherwise, another partition is generated. Eventually, the algorithm terminates since already checked partitions are stored in order to avoid loops.

8.2.3 SAT-Based Partition Search

Due to the vast number of possible partitions, enumerating all of them is practically not feasible. Thus, a further method is proposed, which conducts a SAT-based search. This addresses the crucial task of determining a *good* subset of promising partitions. The basic idea of the SAT-based search $\mathcal{P}_{\text{SAT}}$ is that FFs of a set P forms a *good* partition if (a) an assignment of the fan-in cone exists such that an occurring transient fault in one $FF \in P$ is propagated and visible towards at least one PO and (b) all FFs of P assume the same value. Counting the total number of those test patterns (cf. Theorem 8.3) can be assumed as a metric as formally given in Theorem 8.4 for the qualitative evaluation. Thus, it is possible to rate different sets of FFs (partitions), i.e., the most promising candidate can be derived from this evaluation.

Definition 8.3 Given a set of non-robust FFs $P \subseteq N_L$ of a level L, a test pattern is a set of assignments of PIs such that a transient fault of any flip flop of P is observable at the POs. All test patterns are denoted as $\mathsf{TP}(P)$.

Definition 8.4 Given a set of test patterns of FFs $\mathsf{TP}(P)$. The test pattern metric (denoted $\mathcal{M}(P)$ for a given partition P) is defined as

$$\mathcal{M}(P) = \frac{|\mathsf{TP}(P)|}{2^{|X|}}$$

where X defines the number of PIs as well as FFs in the fan-in cone of partition P.

Observation 1 *Given two sets of FFs $P_1 \subseteq N_i$ and $P_2 \subseteq N_i$ of a circuit level i and let $\mathcal{M}(P_1)$ and $\mathcal{M}(P_2)$ be two ratings of sets of FFs P_1 and P_2, respectively. Then, without loss of generality, $\mathcal{M}(P_1) > \mathcal{M}(P_2)$ states that P_1 is more suited to*

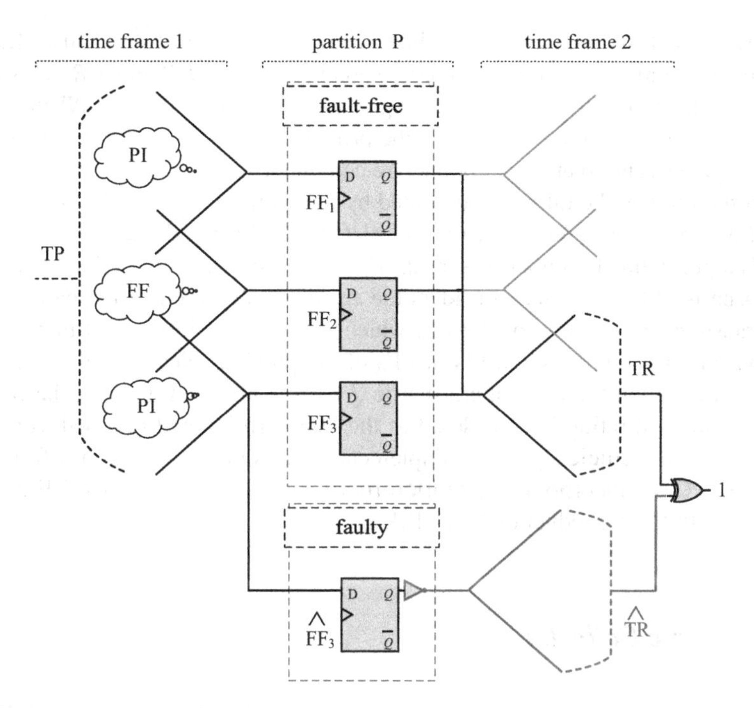

Fig. 8.2 SAT-based ATPG-inspired circuit model

satisfy an EP *than* P_2. *This is due to the relatively high percentage of states, which are considerable for an* EP.

Formal Model
This paragraph describes the underlying algorithmic approach in more detail, which takes advantages of the Observation 1 to, eventually, implement the proposed SAT-based partition search. The proposed method is inspired by ATPG, particularly, the formal model typically used by SAT-based ATPG [Lar92, Dre+09, EWD13] as introduced in Sect. 3.4.

Given a set P of FFs that needs to be rated. The basic principle of the SAT-based approach is shown in Fig. 8.2 for partition P, more precisely, the adopted model of the SAT-based ATPG [EWD13, DS07] is shown. This model is used to generate the SAT-instance and mainly consists of the following: At first, the fault-free sub-circuit, which behaves like the original circuit. All FFs $FF_1, \ldots, FF_3$ of partition P are included, whose outputs are assumed to be equal. Additionally, the fan-in cone of all these FFs are included in the sub-circuit as well, i.e., towards the previous hierarchical FF level or towards the PIs, respectively. Second, the faulty sub-circuit holds a single $\widehat{FF}_3 \in \mathcal{P}$ modeling a transient fault. Hereby, the sub-circuit contains the fan-in and fan-out cone from the fault-free as well as faulty sub-circuit are driven by the same Test Pattern (TP). It is assumed that this transient fault—while following the fault injection of [Fey+11]—tampers at least one signal in the cone of Test Response (TR), which is directly driven by $\widehat{FF}_3$. Thus, a Boolean difference is enforced between the corresponding signals TR and $\widehat{TR}$. This model is translated into SAT-problem resulting in a SAT-instance. If the SAT-instance is satisfiable, there is an assignment of the primary inputs—forming the test pattern. All test patterns up to an upper bound are computed.

To summarize, the rating is computed by calculating $\mathcal{M}(P)$ as follows: Given a set of non-robust FFs N_L of a circuit level L. The SAT-based search first randomly builds a pre-defined number of subsets $P_1, \ldots, P_n$ of N_L. A combinational ATPG approach is utilized, which considers the spanned sub-circuit of the partition, i.e., the transitive fan-in cone of the FF, which is contained in the partition. Hereby, the ATPG considers only the PIs or FFs of the previous hierarchical circuit level. For each partition the test pattern metric $\mathcal{M}(P)$ is obtained. Finally, the partition with the highest rating is considered as the best partition and returned as a result. To improve the efficiency of the implementation, common techniques from SAT-based ATPG are incorporated like Incremental Satisfiability [DS07, TED10] and Cone-of-Influence reduction [Loi+13].

8.2.4 *State Collector*

Once promising partitions have been determined by the Partition Enumerator, the main task of the State Collector is to determine reachable states $\widehat{S}$ such that the EP holds for a partition P_j, which has been obtained by the Partition Enumerator. These states are collected by orchestrating the BMC as introduced in Sect. 3.5. The BMC formulation is revised to determine a path of states so that, eventually, the EP holds for the currently considered partition P_j.

More precisely,

$$\mathsf{SFind}(P_j, l) = I(s_0) \bullet \bigwedge_{0 \leq i < l} T(s_i, s_{i+1}) \bullet \mathcal{P}$$

is employed where $\mathcal{P}$ is a logical formula modeling $\mathsf{EP}(s_l, P_j)$. The formula SFind is satisfiable if there is at least one path $s_0 \ldots s_l$ such that all FFs in the currently considered partition P_j assume the same output value at state s_l. The number l of transitions to be considered, i.e., the unrolling depth of the circuit, can be iteratively increased until either a state s_l has been determined or the maximum unrolling depth (defined by the designer) is reached. The selection of the maximum unrolling depth l while keeping run-time reasonable is a challenging task. To cover the entire search space, l needs to be aligned with the completeness threshold as stated in Algorithm 6. However, from a practical point of view, this is often unfeasible and, moreover, (mostly) not required at all. For many practical instances, getting a

Algorithm 6 State Collecting procedure

1: $\widehat{S} = \emptyset$ {stored as BDD}
2: **for** $k = 1$ to l **do**
3: $F = \mathsf{SFind}(P_j, k)$
4: **while** $\mathsf{SAT}(F)$ **do**
5: **if** $|\widehat{S}| > u$ **then**
6: **return** $\widehat{S}$
7: **else**
8: $\widehat{S} = \widehat{S} \cup s_{i+1}$ {collects state}
9: $F = F \bullet \neg s_{i+1}$ {blocks solution}
10: **end if**
11: **end while**
12: **end for**
13: **return** $\widehat{S}$

satisfactory result is possible without reaching the completeness threshold.[1] If none of such a path can be determined, the partition P_j has been found unsuitable since no state could have been determined in which all FFs in P_j assume the same output value. A further parameter $u > 0$ is added, which prevents the State Collector from determining too many states. Considering too many states increases the complexity of the FDM significantly while hardly improving the achieved robustness anymore.

Algorithm 6 describes the State Collecting algorithm, which receives the partition P_j from the Partition Enumerator, the maximum number u of states to be generated as well as the unrolling depth l for the underlying BMC problem. The collected states $\widehat{S}$ are compactly represented by means of BDDs [Bry86] and initialized by the empty set in line 1. As long as the maximum number u of states to be determined is not reached (line 6), further states are computed. This is achieved by formulating the BMC problem (line 3) for the currently considered unrolling depth k $(0 < k < l)$. Afterwards, the resulting formulation (denoted by F) is solved by a SAT solver (line 4). As long as a satisfying solution is determined, the corresponding states are added to $\widehat{S}$ (line 8) and, afterwards, blocked in the BMC formulation F so that new states can be determined (line 9).

Example 8.5 Given the exemplary partition $P_j = \{FF_3, FF_4, FF_5\}$ as provided by the Partition Enumerator, states shall be determined so that all FFs in P_j assume the same output value. Assuming that both cones f_1 and f_2 in the circuit from Fig. 8.1 do not contain any FFs, a State Collector instance according to SFind is formulated. Solving this instance yields a satisfying solution where all FFs $\{FF_3, FF_4, FF_5\}$ have the same output value "0." From that, it is shown that $\mathsf{EP}(\widehat{S}, P_j)$ holds under the state $s \in \widehat{S}$ defined by $FF_1 = 0$ and $FF_2 = 0$. Consequently, the partition P_j is valid. In the Partition Enumerator (Algorithm 5), this EP is later stored in the container $\mathcal{Z}$.

[1] Note that there are many techniques available, which are approximating the reachable state space of digital circuits and, hence, those techniques can significantly improve the run-time [Gru+05, McM03, RS95]. However, they are not considered in this work to keep the presentation clear.

8.2.5 Fault Detection Mechanism

The previously described methods yield a data container $\mathcal{Z}$ including all application-specific knowledge, which has been obtained in terms of EP*s*, i.e., all valid partitions P_j and corresponding states $\widehat{S}$ that satisfy the EP. This knowledge is now utilized in order to synthesize an FDM. More precisely, for each determined $\mathsf{EP}(\widehat{S}, P_j) \in \mathcal{Z}$, a fault signal $\mathcal{F}$ is to be generated, which is set to "1" whenever the circuit is in a state $s \in \widehat{S}$ (checked by the activator) and, at the same time, the FFs in the partition P_j do not assume the same value (checked by the Equality Comparator). As initially mentioned, the proposed approach allows detecting multiple transient faults if they occur in FFs of disjunctive partitions. In this case, each fault is detected by a separate FDM and, hence, it is excluded by construction that these faults mask each other. In the following, details on the realization of this FDM are provided.

Activator $\mathcal{A}$
For a given $\mathsf{EP}(\widehat{S}, P_j)$, a signal $\mathcal{A}$ has to be created which is set to "1" if the circuit is in a state $s \in \widehat{S}$. All currently relevant states $\widehat{S}$ are stored in terms of a BDD and, hence, corresponding logic triggering signal $\mathcal{A}$ can easily be derived from the BDD (see Sect. 3.7) by replacing all nodes with a corresponding multiplexer gate like [DSF04].

Besides that, the timing of $\mathcal{A}$ has to be adjusted appropriately. Transient faults are assumed to occur in the transition between two consecutive states. This is why the states $s \in \widehat{S}$ are collected for state s_{l-1} (assuming their effects manifest in state s_l). Consequently, the check for states has to be conducted one state before the values of all FFs in P_j are to be compared, i.e., $\mathcal{A}$ has to be generated one state before the comparison is conducted. This requires $\mathcal{A}$ to be buffered for one cycle, which is accomplished by introducing an additional FF L-Act_1. However, since L-Act_1 is vulnerable to transient faults, robustness is not guaranteed anymore. Thus, a second FF L-Act_2 is introduced, which also receives the value of signal $\mathcal{A}$. After one cycle, the output values of both FFs L-Act_1 and L-Act_2 are checked for equivalence. If the two values are not equal, a fault is reported (by setting the fault signal $\mathcal{F}$ to "1").

Example 8.6 Consider the running example again with the circuit from Fig. 8.1 and the determined $\mathsf{EP}(\widehat{S}, P_j)$. Figure 8.3 shows the resulting circuit created by the proposed FDM scheme. The bottom left corner of Fig. 8.3 sketches the resulting Activator logic. More precisely, using $\widehat{S}$ represented as BDD [Lee59], a multiplexer circuit is created which generates the signal $\mathcal{A}$ (sketched by the block State Collector). Then, the resulting signal is passed to the two FFs L-Act_1 and L-Act_2. To protect both newly inserted FFs against possible single transient faults, an XOR gate is introduced. In case of a deviation between both FFs L-Act_1 as well as L-Act_2, the Activator Fault signal is triggered, which, especially, overdrives the fault signal $\mathcal{F}$ to indicate an occurred fault.

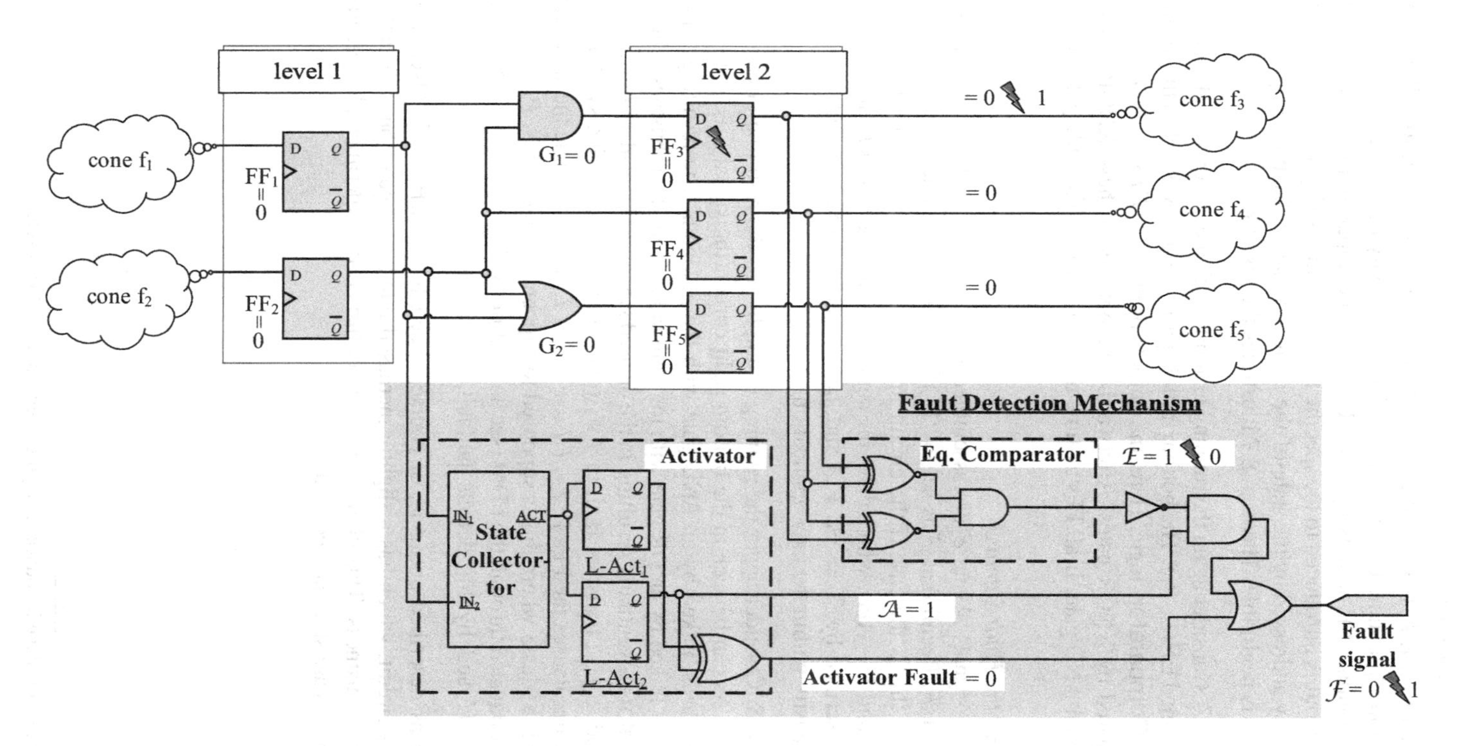

Fig. 8.3 Applying the proposed methodology to the circuit from Fig. 8.1

Equality Comparator $\mathcal{E}$

For a given EP($\widehat{S}$, P_j), a signal $\mathcal{E}$ has to be created which is set to "1" if the FFs in a partition P_j assume the same values. This can easily be realized by connecting the corresponding FF outputs by XNOR gates and comparing their result. Since a further fan-out is introduced to the specific FF, the timing is slightly affected, which can be easily addressed by the state-of-the-art techniques of the power optimization and, hence, is negligible. Theorem 8.7 illustrates the resulting logic.

Example 8.7 Consider again the running example with the circuit from Fig. 8.1 and the determined EP($\widehat{S}$, P_j). The bottom middle part of Fig. 8.3 sketches the resulting Equality Comparator logic. Here, the outputs of all FFs $\{FF_3, FF_4, FF_5\} \in P_j$ are compared by XNOR gates. Afterwards, the outputs of these XNOR gates are passed to an AND gate. If all FFs assume the same value, this AND gate evaluates to "1."

Generating the Fault Signal $\mathcal{F}$

Finally, the signals $\mathcal{A}$ and $\mathcal{E}$ are assembled into a single FDM that generates the fault signal $\mathcal{F}$. Recall that $\mathcal{F}$ is set to "1" if a fault has been detected. For a given EP($\widehat{S}$, P_j), this is the case if the circuit just left a state $\widehat{S}$ (stored in the both FFs $L - Act_1$ and $L - Act_2$, which drive the signal $\mathcal{A}$)[2] and all FFs in $\mathcal{P}_j$ are not equal. This is described by $\mathcal{F} = \neg\mathcal{E} \bullet \mathcal{A}$ and, hence, an occurred transient fault can be detected easily within the later application.

Example 8.8 Consider again the running example and the resulting circuit shown in Fig. 8.3. As can be seen in the bottom right corner, the signal $\mathcal{E}$ is first inverted and, afterwards, linked by an AND gate with the $\mathcal{A}$ signal. The resulting value is additionally linked by an OR gate with the fault value from the robustness check of the Activator—eventually resulting in the desired signal $\mathcal{F}$. Consider now the entire circuit and, e.g., a transient fault in FF_3 (denoted by the red strike symbol). This circumstance causes the FFs $\{FF_3, FF_4, FF_5\} \in P_j$ to not assume the same value anymore in states $\widehat{S}$ where this is supposed to happen, i.e., the EP($\widehat{S}$, P_j) fails. This case is propagated through the FDM (see annotations in Fig. 8.3), which, eventually, sets $\mathcal{F}$ to "1," and, by this, detects the fault.

Logic as described above is, of course, added for all EP $\in \mathcal{Z}$. Overall, this leads to a circuit, which has a slightly increased number of gates but substantially improved robustness. This claim has been confirmed by experimental evaluations whose results are summarized next.

[2] Note that this FF stage introduces a required delay of one clock cycle.

8.3 Experimental Setup

This section describes the experimental setup of the proposed knowledge-based transient fault detection methodology, which has been completely implemented in C++. This framework includes a simulation-based robustness checker, which determines the non-robust FFs of a given circuit by transforming the circuit—provided in Verilog—into a compiled simulation model. More precisely, the Verific's [Inc20] parser platform is invoked to parse the Verilog code and to generate an intermediate program (IR Code)by utilizing the Low Level Virtual Machine (LLVM) [LA04].

The SAT solver MiniSAT [ES04] on top of metaSMT [Rie+16] in combination with the X-value abstraction technique [Gru09] has been utilized to conduct the particular BMC instance and the SAT-based partitioning approach. The BDD package CUDD [Som16] has been used to generate the multiplexer circuits. The resulting flow has been evaluated using ITC'99 benchmark circuits. In order to determine the set of all non-robust FFs, the parameters $l = 500$ and $r = 5$ for simulation have been found suitable for these circuits according to the (minimal) latency analysis of [FSF14]. The Partition Enumerator considered different maximal partition sizes holding maximal $p_s \in \{8, 16\}$ FFs.

The State Collector always assumed an unrolling depth of $l = 10$ and a bounded number $u = 1024$ of states to be collected per partition P_j.[3] The selection of the unrolling depth l is aligned with the maximal latency, which is determined for the simulation-based approach in work [FSF14] and acts as an upper bound for the applied formal BMC model. The developed flow seamlessly integrates both the random as well as the SAT-based technique. This allows the designer to select a trade-off easily between the overall robustness improvement with respect to the introduced hardware overhead.

All evaluations have been conducted on a Fedora 30 (with gcc 9.2.1) system holding an *Intel Xeon E5-2640v4 2.4 GHz* processor with 256GB system memory. The TO is set to 96h and the MO is set to 64GB.

8.4 Experimental Results

This section describes the obtained results of the experimental evaluation in detail, which are summarized as follows:

- Table 8.1 provides details on the considered benchmark circuits, i.e., its circuit name (**circuit**), the number of gates (**#gates**), and the number of FFs (**#FFs**), as well as the **run-time** (in CPU minutes) required by both the random $\mathcal{P}_{\mathsf{RAND}}$ and the SAT-based $\mathcal{P}_{\mathsf{RAND}}$ when maximal partition sizes of $p_s \in \{8, 16\}$ are applied.

[3]Note that the bound u was not exceeded.

Table 8.1 Run-time and FDM sizes for different $p_s \in \{8, 16\}$

			$\mathcal{P}_{\text{RAND}}$ run-time [m]		$\mathcal{P}_{\text{SAT}}$ run-time [m]	
Circuit	#gates	#FFs	$p_s = 8$	$p_s = 16$	$p_s = 8$	$p_s = 16$
b05	608	66	< 0.10	< 0.10	< 0.10	0.15
b06	66	9	< 0.10	< 0.10	< 0.10	< 0.10
b07	382	51	0.18	0.18	2.96	44.03
b08	168	21	< 0.10	< 0.10	< 0.10	< 0.10
b09	131	28	< 0.10	< 0.10	2.88	11.25
b10	172	17	< 0.10	< 0.10	0.26	17.18
b11	366	30	< 0.10	< 0.10	1.50	0.28
b12	1000	121	1.26	1.15	0.71	0.911
b13	309	53	< 0.10	0.11	21.98	791.67

Table 8.2 FDM stats for different $p_s \in \{8, 16\}$

	$\mathcal{P}_{\text{SAT}}$ #FDMs		$\mathcal{P}_{\text{SAT}}$ #Nodes	
Circuit	$p_s = 8$	$p_s = 16$	$p_s = 8$	$p_s = 16$
b05	1	1	5	4
b06	1	1	5	4
b07	6	3	59	16
b08	1	1	13	13
b09	2	1	43	6
b10	2	1	11	3
b11	3	2	21	10
b12	1	1	30	32
b13	3	2	9	6

- Table 8.2 gives in column $\mathcal{P}_{\text{SAT}}$ **#FDM** the overall number of introduced FDMs into the circuit by applying the SAT-based approach for both maximal partition sizes. Furthermore, the column $\mathcal{P}_{\text{SAT}}$ **#Nodes** shows the overall nodes of *all* state collectors. Note that each FDM holds exactly one individual State Collector, which is represented as a BDD and, hence, can be synthesized by a one-to-one mapping between nodes and multiplexer gates as stated in [DSF04].
- Fig. 8.4 shows the hardware overhead (in terms of a gate count in percentage) caused by applying the proposed methodologies, i.e., one random-based as well as SAT-based approach, for the considered maximal partition sizes.[4]
- Fig. 8.5 shows the robustness of the original circuit as well as the robustness after applying both proposed methodologies (again for different maximal partition sizes $p_s \in \{8, 16\}$).
- Fig. 8.6 presents a direct comparison between the random-based and SAT-based approaches: The Scale value on the X-axis shows the difference $\mathcal{H}_{SG} - \mathcal{H}_{RB}$ of the random-based scale factor $\mathcal{H}_{RB}$ and of the SAT-based scale factor $\mathcal{H}_{SG}$. Thus, the lower the *scale* bar chart is, the lower the introduced hardware overhead

[4] If no bar is shown, a hardware overhead of 0% or close to 0% is measured.

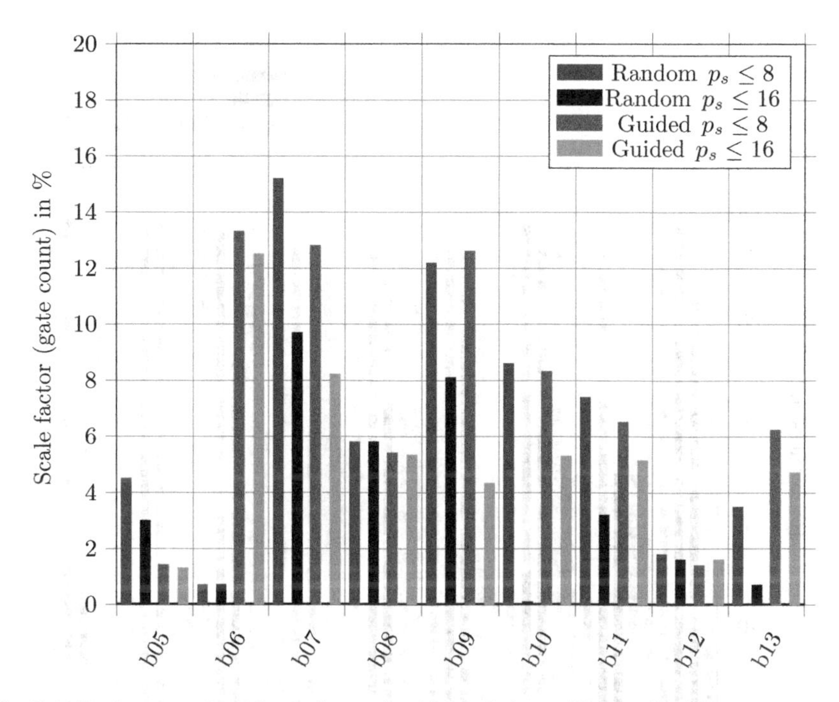

Fig. 8.4 Hardware overhead for random and guided technique with $p_s \leq \{8, 16\}$

is while applying the SAT-based approach instead of the random-based one and vice versa. The resulting robustness improvement of the enhanced designs (by utilizing the proposed approaches) $\mathcal{R}_{RB}$ and $\mathcal{R}_{AG}$ are compared with respect to the design's initial robustness[5] $\mathcal{R}_{init}$ is determined as follows: $\Delta\mathcal{R}_{RB} = \mathcal{R}_{RB} - \mathcal{R}_{init}$ and $\Delta\mathcal{R}_{SG} = \mathcal{R}_{SG} - \mathcal{R}_{init}$, respectively. The Robustness on X-axis provides the difference of both robustness improvements: $\Delta\mathcal{R}_{SG} - \Delta\mathcal{R}_{RB}$. Thus, the higher the robustness bar chart is, the higher the robustness improvement is, which is achieved while applying the SAT-based approach instead of the random-based one and vice versa.

The results nicely show the effect of different partition sizes p_s as stated in Fig. 8.4: In almost all cases, a larger p_s leads to a smaller hardware overhead. This is because larger partitions cover more FFs and, hence, require the consideration of a smaller total number of partitions leading to less FDM logic. This observation is both valid for the random-based as well as the SAT-based approach. Figure 8.6 presents a direct comparison between the random-based and the SAT-based approach. The SAT-based approach allows reducing the introduced hardware overhead in two-third of the conducted experiments while achieving at least the

[5]Note that the initial robustness is due to the circuit's structure and, more precisely, due to the implicitly given redundancies as observed in [TD09].

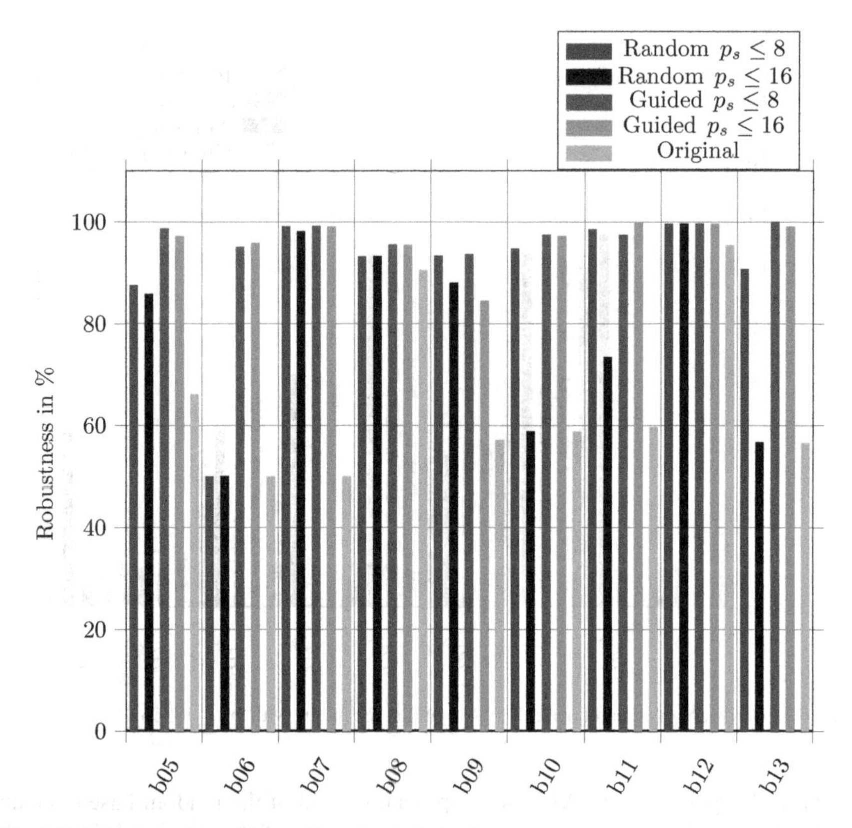

Fig. 8.5 Robustness improvement for enhanced circuits

same (or even a slightly higher) robustness enhancement. In the case of benchmark circuits b06 (b13), the newly proposed SAT-based approach clearly shows its advantages against the random-based one: The robustness is even more increased by approx. 55% (10%). This observation can also be validated for the SAT-based approach by considering Table 8.1: The number of introduced FDMs is lower equal when invoking the SAT-based technique with $p_s \leq 8$ instead of $p_s \leq 16$. For instance, we see 3 (2) FDMs being introduced to the circuit b11 with p_s equals to 8 (16). This leads to overall 21 (10) nodes in the generated state collectors. At the same time, this reduces the required run-time since fewer BMC checks have to be conducted. A significant share of the overall run-time is caused by the collection of the states and, more precisely, the storage within a BDD structure. Both methodologies are slightly able to improve the robustness for b08 only. Moreover, the hardware overhead is even the same.

The proposed methodology provides a suitable alternative to previously proposed solutions, as initially discussed. The space-based approaches such as Triple Modular Redundancy [BW89, SB89] can guarantee 100% robustness, and they usually require more than thrice the amount of hardware (i.e., yielding a scaling

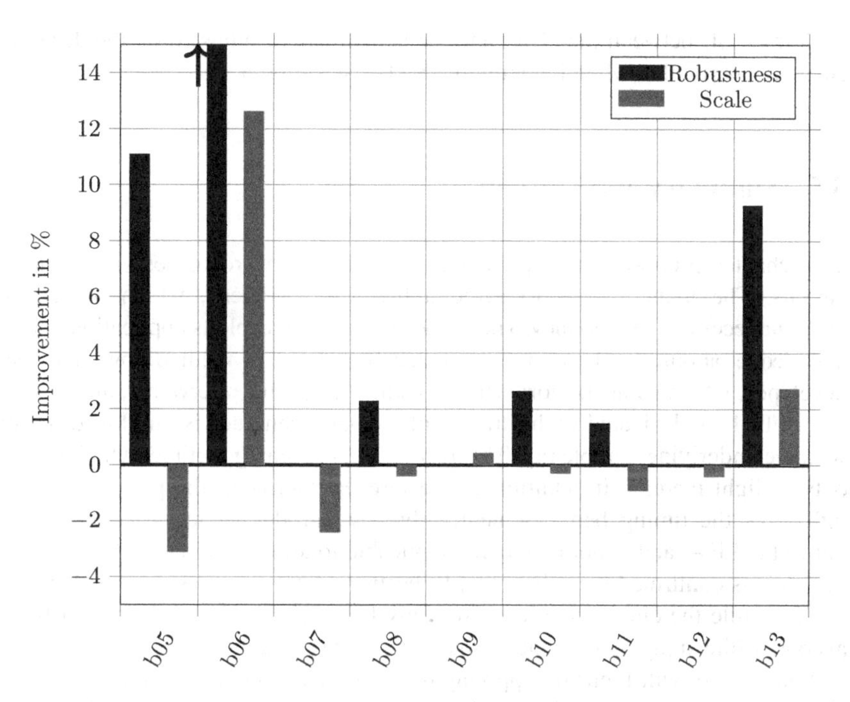

Fig. 8.6 Comparison between random-based and SAT-based approaches

factor of > 3.0). In contrast, the solution proposed in this work is capable of always improving the robustness to more than 90% (in some cases even close to 100%), while only 5.8% more hardware is required for this in average. As already discussed before, the proposed solution also outperforms timing-based and application-specific approaches, since timing is hardly affected at all in the proposed solution and the methodology can be applied to arbitrary sequential circuits. By this, a suitable trade-off between enhancing the robustness and keeping the hardware overhead small is achieved.

The determination of *good* partition is a computationally hard task: While considering structural information as done by the SAT-based approach, the run-time is increased. The run-times are manageable, even though common ATPG techniques, for instance, modeling J-frontiers and D-chain-based propagation, have not been applied. The proposed SAT-based approach tackles this challenge by exploiting the powerful formal model, which is inspired by SAT-based ATPG techniques, leading to:

1. Enhanced robustness in most of the cases compared to the random-based approach while the robustness has not been reduced in any considered cases and
2. the scale factor is in almost all cases reduced compared to the random-based approach or the robustness has been significantly enhanced.

A trade-off between the effort and the benefit can be adjusted on the designer's choice, in particular, when invoking the SAT-based approach.

8.5 Summary and Outlook

This chapter proposed an approach for improving the robustness of sequential circuits. The main idea is to avoid a hardware overhead, which often introduces unnecessary redundancy. The proposed approach exploits application-specific knowledge about the FFs in each reachable state. A methodology has been developed, which gains the corresponding knowledge and, afterwards, utilizes them for a FDM. A dedicated orchestration of formal techniques is employed to cope with the underlying complexity. This results in a hardening method, which requires only a slight increase in additional hardware. Furthermore, the proposed method influences the timing behavior negligibly—by introducing just one further fan-out to the FFs—and is automatically applicable to arbitrary circuits. Experimental evaluations confirmed these benefits: Robustness can be increased to approx. 84% (97%), while the circuit size increases only by a factor of approx. 1.07 (1.07) on average while applying the random-based (SAT-based) approach.

Future work will focus on applying the proposed technique to large (industrial-sized) circuits. One method towards this concerns the consideration of a compositional approach. Such a compositional approach is often applied to reduce the search space when employing formal methods and follows the idea of selecting parts of the circuits and process these parts individually. In comparison to the monolithic approach, the compositional one allows highly reducing the resulting search space since manageable parts are hardened separately. Furthermore, an alternative approach for the determination of the most beneficial partitions should be exploited by developing an application-specific evolutionary algorithm coupled with a hardware-accelerated circuit emulator by using field programmable gate arrays. The hardware-accelerated approach allows investigating a comparatively large part of the state space of a given sequential circuit when determining beneficial candidates (partitions). It is assumed that the quality—in the sense of the later robustness enhancement—can be further improved in conjunction with a reduction of the overall run-time of the hardening procedure since the time-consuming partition enumeration and evaluation is being outsourced. The determined partition data are then fed back into the regular flow as presented in this chapter to, finally, invoke the BMC-based synthesis of the EP and to generate the design with the enhanced robustness.

Chapter 9
Conclusion and Outlook

The integration of *Design for Testability* (DFT) measures is strictly required when designing state-of-the-art *Integrated Circuits* (ICs) to, among others, ensure that a *good* testability prevails in the resulting design. This testability allows performing high quality manufacturing tests, which give a certain level of confidence that no defects have occurred during the manufacturing process, which potentially tamper the correctness of the functional behavior. However, the steadily increasing design complexity yields a significant increase in the *Test Data Volume* (TDV) and, hence, the *Test Application Time* (TAT), which both increase the test costs. This effect is even more amplified when generating tests for safety-critical applications like automotive systems, enforcing a zero defect policy.

Different test compression techniques were proposed in the literature, which allow to significantly reduce the TDV during the high volume manufacturing test under certain conditions during the wafer test. However, these techniques are, for instance, not accessible during the *Low-Pin Count Test* (LPCT). Thus, a high TAT results or the specific tests are even not applicable due to exceeded (tester) memory resources. In addition to these testability criteria, further requirements have to be met, for instance, the overall number of input and output pins at the chip-level, whose number heavily affects the resulting manufacturing costs. Thus, a contradictory objective between the testability and the test as well as manufacturing costs exists and for which a trade-off has to be determined. Analogously to the DFT measures, specific structures in the sense of *Design for Debug and Diagnosis* (DFD) are introduced since similar problems exist when debugging complex systems.

This book tackles the aforementioned challenges to pave the way for the next generation IC design. In particular, measures were developed and implemented, which enable a significant reduction of the TDV as well as the TAT and, hence, to massively reduce the resulting test costs. One major contribution of this book comprises *Test Vector Transmitting using enhanced compression-based TAP compression-based TAP controllers* (VecTHOR):

S. Huhn, R. Drechsler, *Design for Testability, Debug and Reliability*,
https://doi.org/10.1007/978-3-030-69209-4_9

VecTHOR implements a codeword-based compression technique, which combines a dynamically configurable dictionary with a run-length encoding scheme. The combination of these techniques had been proven as very effective to process homogeneous as well as heterogeneous test data fractions. The compression technique was seamlessly integrated into an IEEE 1149.1 shell, which yielded a lightweight TAP controller with embedded compression, fully compliant with the frequently used IEEE 1149.1. Thus, VecTHOR is meant to be integrated into a state-of-the-art design, which employs *Joint Test Action Group* (JTAG) controller, in any case, to take advantage of a TDV reduction of 50% in any test scenario.

Alongside the development of VecTHOR, a corresponding retargeting framework was implemented, which introduced formal optimization *Boolean Satisfiability* (SAT)-based techniques to generate the compressed test data off-chip once and to determine an optimal set of codewords. In contrast to the recently used greedy approach, the optimization-based technique allowed to improve the TAT reduction results significantly, in particular when processing hard-to-compress random test data. An elaborated partitioning scheme was developed, which considered, among others, the current configuration state of the embedded dictionary. This partition-based approach enables the processing of a large set of test data from industrial representative use-cases by formal optimization-based techniques. By this, a speed-up of approx. 18x was achieved compared to the monolithic optimization-based approach without any partitioning being applied.

Furthermore, a hybrid compression derivative was designed out of VecTHOR, which has been done in tight cooperation with Infineon Germany. This hybrid approach addresses the shortcomings of existing compression techniques specifically in the field of low-pin count test environments as frequently given in the field of zero defect testing targeting automotive applications. Moreover, the hybrid approach revealed the untapped potential of unused multichannel topologies of the IC designs and, hence, the resulting test data were reduced by up to 74%, which lowers the resulting costs by 3x. All developed techniques were extensively tested by using industrial and high-entropic data and were verified by simulation and synthesis and reached to a level, where they can potentially be deployed to an industrial environment for a prototypical integration in a test chip.

Formal techniques were further applied in Chap. 8 to *harden* the IC design, i.e., to reduce the vulnerability against transient faults. Such a transient fault occurs in harsh environments like a high level of radiation, which (potentially) invalidate the functional behavior. Analogously to faults induced by manufacturing defects, a transient fault occurring during the functional operation can lead to the same disastrous consequences, particularly when being involved in a safety-critical application scenario. To achieve this, application-specific knowledge was exploited about the structural dependencies between *Flip-Flops* (FFs) with respect to their reachable states. This knowledge was used to deduce equivalence properties—a newly developed formal concept to describe these structural dependencies. To this end, a methodology was introduced, which gained the corresponding knowledge and, afterwards, utilized the knowledge to synthesize certain *Fault Detection Mechanism* (FDM) structures. A dedicated orchestration of formal techniques was employed, yielding a hardening method for arbitrary sequential circuits. This technique introduced only a slight hardware overhead, which was about 7% at maximum. However, the robustness was increased by up to 97%.

Future work aims at extending the proposed hardening technique of Chap. 8 such that a transient fault can be directly corrected in addition to its detection. It is required to store the condition of the corresponding *Equivalence Property* within the StateCollector to implement an extended approach, which allows the correction of a recently occurred single transient fault within two subsequent clock events before the fault value induces, for instance, a silent data corruption. In the field of DFT, future work will focus on a seamless integration of VecTHOR within the *Internal Joint Test Action Group* (IJTAG) [14] on different levels. For instance, VecTHOR can act as a user-defined wrapper for an arbitrary instrument, serve as the chip-level interface of the whole network or can be used to compress the configuration data of the network exclusively. Moreover, these levels can be combined to form a multi-stage compression architecture, which will improve the compression even more. In particular, it is expected that the integration within the chip-level IJTAG interface can be conducted without introducing a large hardware overhead since this chip-level access is typically realized by a state-based access mechanism like IEEE 1149.1. By this, the overall application time of an IJTAG network would be significantly reduced by at least 3x by integrating a compression-based access mechanism. This reduction would allow entirely new test applications since the reconfiguration of an individual scan path would be massively accelerated as well. The described expectation is founded on the already achieved compression ratio of approx. 60% for industrial representative test data, which can be further increased since the configuration data includes repeating sequences, which are most suitable for being addressed by VecTHOR.

Appendix A

This appendix presents the seven approaches that have been developed in the context of VecTHOR's retarget framework. The authors make these developments publicly available at

https://unihb.eu/VecTHOR

under the terms of the MIT license. These approaches include four fast heuristic methods that invoke a word-based Huffman encoding technique and, furthermore, three formal approaches using a PBO solving technique. Since the basic principle of these retargeting techniques have been described in Sect. 4.3 (on page 66) and in Sect. 5.2 (on page 78), this appendix focuses on the software-side of the developed framework.

A.1 Overview of Retargeting Techniques

This section describes the presented retargeting techniques in more detail and discusses the expected results when applied to different benchmarks. The individual approach can be enabled by setting the configuration accordingly, which is further described in the next section.

Four different heuristic retargeting techniques have been developed. These techniques allow a fast retargeting completion even for large incoming data. Depending on the applied technique, a promising TDV reduction ratio is achieved. In contrast to this, these techniques typically introduce an overhead concerning the TAT. More precisely, the Compress technique executes a heuristic approach while solely considering the pre-configured embedded dictionary without any reconfiguration. As a reference, the Merge-Compress technique introduces a run-length encoding capability to the retargeting procedure, which allows merging long series of same datawords effectively. The Dynamic-Compress technique takes advantage of the dynamic reconfiguration of the embedded dictionary, which improves the

S. Huhn, R. Drechsler, *Design for Testability, Debug and Reliability*,
https://doi.org/10.1007/978-3-030-69209-4

compression efficacy significantly. Finally, Dynamic-Merge-Compress combines both the dynamic reconfiguration with the run-length encoding capability and, by this, achieves the best results from the heuristic approach with respect to the reduction of the TDV and TAT, respectively.

1. Compress
2. Merge-Compress
3. Dynamic-Compress
4. Dynamic-Merge-Compress

Besides the heuristic approaches, different techniques have been proposed invoking formal techniques. More precisely, the task of determining a most beneficial set of codewords has been mapped to a PBO instance, yielding the three approaches as follows:

6. SAT-Compress
7. SAT-Postprocess-Compress
8. Partition-SAT-Compress

The SAT-Compress invokes the PBO-solver with the generated objective function, as described in Sect. 5.2.2 (on page 80). In comparison to the heuristic approach, the SAT-based retarget procedure achieves a higher TDV and TAT ratio. However, the required computing time is significantly higher, which limits the applicability to a large data volume. In addition to SAT-Compress, the SAT-Postprocess-Compress introduces a second optimization stage, aiming to reduce the TAT even stronger, which requires more computing time. To allow the processing of even large data volume, the Partition-SAT-Compress approach introduces a partition scheme on top of the PBO instance. By this, the computing time can be significantly reduced while retaining nearly the same TDV and TAT reduction ratios.

A.2 Retargeting Framework

This section describes the retargeting framework. At first, the required prerequisites are discussed, followed by a presentation of the available parameters to adjust the retargeting procedure.

A.2.1 Getting Started

The following paragraph briefly describes the prerequisites for getting started with VecTHOR and explains the available options. Since VecTHOR is solely implemented in C++ and utilizes heavily C++11 features, a compatible C++ compiler version has to be used. The code has been tested with the GNU Compiler Collection gcc v.4.9 or newer. Besides the compiler, the boost C++ Library [Kar05] is

required with a version compatible to version 1.41.0 and the cross-building platform CMake in version 2.8 or newer. The framework orchestrates clasp 3.1.4 [Geb+07] as the underlying PBO solving engine and YAML [BEI09] to process user-defined configuration files.
Note that the retargeting frame operates in a terminal-only mode. For a future version of VecTHOR, a graphical user interface is planned as an additional feature. After the compilation has been completed, VecTHOR can be executed by calling the compiled and linked executable.

A.2.2 Available Options

Several options exist to control VecTHOR's behavior. This can be either done by using different call parameters or by using the configuration file. If VecTHOR is invoked with the –help parameter, the list of available call parameters will be emitted, as shown in Listing A.1.

```
[user@pc VecTHOR ]$ VecTHOR --Help

    [i]Allowed options:
    --Help                  produces help message
    --Verbose               be verbose
    --Debug                 prints lots of debug info
    --Stats                 prints stats
    --Plot                  generates plots
    --Hex                   processes data as hex
    --STIL                  processes data as STIL format
    --ConfigFile arg        reads configuration file
    --ReadTDR arg           reads external TDR data file
    --NumRTDR arg           number of bytes to be generated (random)
    --LegacyJTAG arg        generates legacy JTAG sequence
    --CompressedJTAG arg    generates compressed JTAG sequence
    --WriteGolden arg       write golden file for comparison
```

Listing A.1 Allowed terminal options

The call parameters are as follows:

Verbose|Debug|Stats Boolean flags to control the log level, i.e., the different level of details being printed to the console. Note that this possibly exceeds your buffer when using the debug log level on large incoming test data.

Plot Boolean flag to control whether VecTHOR creates specific gnuplot files, which are linked to the buffering functionality that has not been described in detail in this book.

Hex Boolean flag to enable the processing of incoming test data (using the ReadTDR parameter) that are encoding in hex.

STIL Boolean flag to enable the processing of incoming test data by using an external STIL parsing library.

ConfigFile <file> Determines the user-defined configuration file that is being processed.

ReadTDR <file> Sets the file that holds the incoming test data.

NumTDR <int> Controls the number of random bits to be generated.

LegacyJTAG <file> Determines the output filename for the corresponding legacy JTAG operation, which can be typically used to measure the reference cycle number precisely and for re-simulation (validation).

CompressesJTAG <file> Defines the filename (and path) to store the compressed test data seamlessly encoded by using the extended JTAG protocol.

WriteGolden <file> Writes the golden test data into a separate file, which can then be used for validation purposes.

As indicated by the parameter ConfigFile above, VecTHOR supports user-defined configuration files, which are encoded in the YAML [BEI09] syntax. The configuration file controls various parameters, which are typically not changed in between two consecutive runs.

```
vecthor:
 merging_repititions : "true"
dynamic : "true"
max_cdws: "12"
# Heuristic only
 heur_inner_freq : "8"
 heur_outer_freq : "4"
heur_weight: "1"
#SAT only
SAT: "true"
SAT_SEC: "true"
SAT_CONFL: "1000000"
SAT_RESTART: "500"
#Log levels
 stats : "true"
verbose: " false "
debug: " false "
# Partition only
 part_size : "8192"
P2S_BUFFER: "false"
hw_emit: " false "
 validate : " false "
 resync_file : " ../ data/resynced.data"
 legacy_prefix : " ../ data/legacy. prefix "
 legacy_suffix : " ../ data/legacy. suffix "
compressed_prefix: " ../ data/dynmergecompressed.prefix"
compressed_suffix: " ../ data/dynmergecompressed.suffix"
dyncompressed_infix: " ../ data/dyncompressed. infix "
dyncompressed_preload: " ../ data/dyncompressed.preload"
```

Listing A.2 Allowed options in configuration file

An exemplary configuration file is shown in Listing A.2, whose parameters are as follows:

merging_repititions Boolean flag to determine whether successive codewords should be merged by taking advantage of the run-length encoding scheme.

dynamic Boolean flag to control whether the embedded dictionary of VecTHOR should be dynamically configured.

max_cdws <int> Controls the (maximum) number of codewords to be configured within the embedded dictionary. Note that this number must be compliant with VecTHOR's settings used for synthesis.

heur_(inner|outer)_freq <int> Two parameters of the heuristic approach for controlling the inner and outer frequency of the nested loops.

heur_weight <int> Determines the step size (in bytes) of the heuristic retargeting procedure. This size should match the maximal datawords length.

SAT Determines whether the SAT-based approach should be invoked. This parameter supersedes the heuristic settings.

SAT_SEC Enables the SAT-based post-processing, which invokes a further optimization step by an incremental call of the PBO-solver.

SAT_CONFL <int> Determines the number of stored conflict (learnt) clauses within the clause database.

SAT_RESTART <int> Defines the maximum allowed number of restarts during the SAT-solving.

Verbose|Debug|Stats Boolean flags to control the log level.

part_size <int> Enables the partition SAT-based approach and assumes the given number as the maximum partition size.

P2S_buffer Enables the buffering approach, which utilizes an additional on-chip buffer such that a continuous data stream is being generated and, hence, no clock-gating is required. Note that this buffer has to be activated during VecTHOR's synthesis, and the size has to be aligned with the retargeting procedure. Further note that an initial delay is introduced during the decompressing to fill the buffer with data.Ât'

hw_emit Boolean flag to activate the hardware-based emitter instead of the simulation-based one. The retargeting framework supports an ARMv8A Cortex-A53 microprocessor and the respective IO shield.

validate Enables the internal validation of the retargeting process.

resync_file <file> Writes further re-synchronization information to the specific file, which allows to re-synchronize the input and output data after invoking the embedded compression technique.

legacy_(prefix|suffix) <file> Files that include the required prefix and suffix regarding the test bench code to execute the simulation-based data transfer via a legacy TAP controller.

compressed_(prefix|suffix) Files that include the required prefix and suffix regarding the test bench code to execute the simulation-based data transfer via the compression-based TAP controller.

dyncompressed_(infix|preload) Files that include the required infix and suffix regarding the test bench code to perform the embedded dictionary's dynamic configuration. Hereby, infix refers to the required signal transitions between the configuration of two codewords and preload to the required instruction code.

A.2.3 Retargeting Procedure

VecTHOR is typically applied to an existing set of test data, which are meant to be retargeted for taking advantage of VecTHOR's significant reduction of the TDV and the TAT. In fact, VecTHOR holds an integrated pseudo-random number generator—based on a Mersenne Twister algorithm—to generate test data of high-entropy. This feature enables a direct evaluation of the used parameter sets on hard-to-compress data, as typically given by data of high-entropy. However, the regular use-case is about calling retargeting directly on a file containing the existing test data. After the retargeting process has been completed, a compressed JTAG file is written out. This emitted file contains a full test bench environment for seamless re-simulation using digital simulation tools. The academic version Questa Sim-64 10.7c has been solely used for this purpose during the development phase of VecTHOR.

A.2.4 Architecture of Framework

This paragraph gives a brief introduction to the underlying software architecture of VecTHOR's retargeting framework. Overall, VecTHOR consists of 21 different classes, which implement the three main components decompressor, compressor, and emitter.

The decompressor realizes the retargeting procedure by determining the most beneficial set of codewords, which is then used to configure VecTHOR's embedded dictionary. The compressor of the retargeting framework then generates the compressed test data stream using an internal, highly efficient bit vector representation. Especially, the emitter generates the compressed data stream either as test bench code for re-simulation (by using the enhanced TAP controller) or by emitting the enhanced TAP controller protocol via a connected IO shield.

Different inheritances while taking advantage of virtual classes like the emitter class allow a straight extensibility to support a further data sink, which is part of the test tool-flow. More detailed collaboration diagrams of the four most important classes are shown in Figs. A.1, A.2, A.3.

Note that these figures follow the same nomenclature as being regularly used in DoxyGen [Lar11] documentation files. More precisely, a filled gray box represents a class for which the collaboration diagram has been generated and a box with a black border denotes a collaborative class. The arrows have the following meaning: A dark blue arrow visualizes a public—protected and private ones are not shown—inheritance relation between two classes and a purple dashed arrow is used if a class is contained or used by another class. All arrows are labeled with the variable(s) through which the collaborative class is accessible.

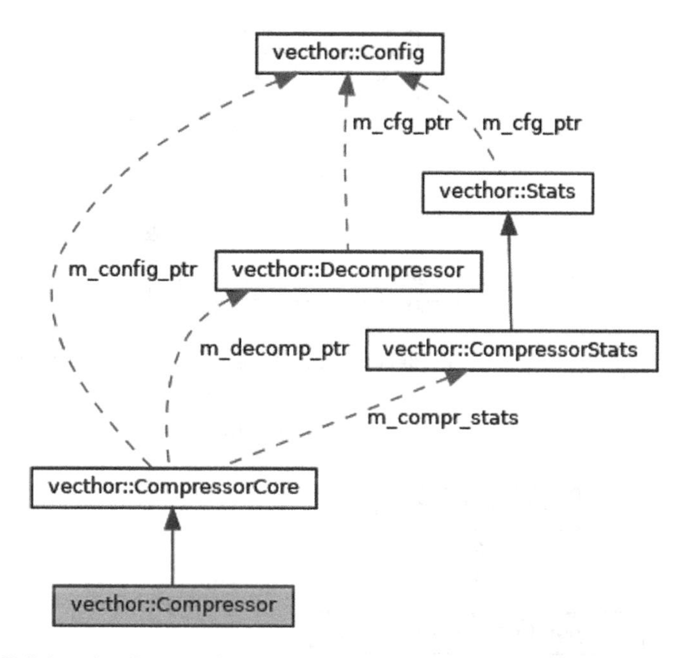

Fig. A.1 Collaboration diagram of compressor class

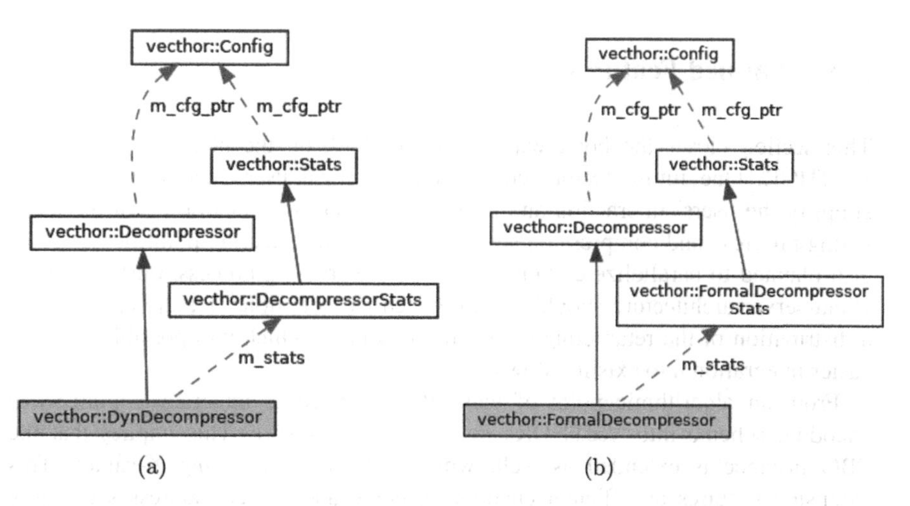

Fig. A.2 Collaboration diagram of inherited decompressor classes. (**a**) DynDecompressor subclass. (**b**) FormalDecompressor subclass

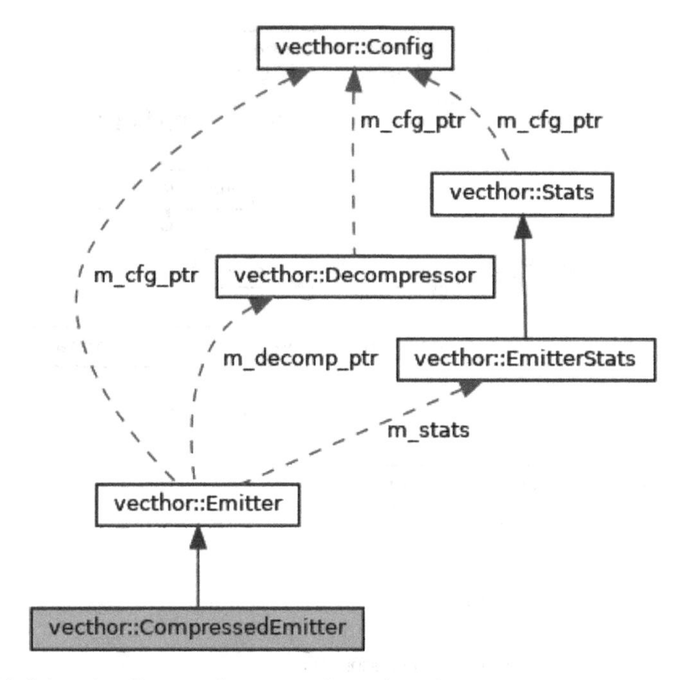

Fig. A.3 Collaboration diagram of representative emitter class

A.3 Planned Features

This section closes the book with a brief outlook on the planned evolution of VecTHOR. One future feature concerns a graphical user interface, which will simplify the users' interaction and provide important information about the retargeting progress and compression success. From an engineering point of view, it is also planned to parallelize certain parts of the retargeting process. Furthermore, a client-server architecture should be introduced as an optional feature allowing for a distribution of the retargeting core and the emitter, which will possibly yield an easier integration into existing flows.

From an algorithmic point of view, it is planned to integrate a multi-value encoding scheme into VecTHOR's retargeting framework. This implies that the PBO instance is extended as well, which will allow modeling X-values. This extension requires an efficient encoding scheme and, hence, addresses a highly relevant research question. It is assumed that this will further increase compression efficacy since some test data are typically unspecified in different test applications. Consequently, the full potential is not yet revealed in such a circumstance.

References

[Ace+17] C. Acero et al., Embedded deterministic test points. IEEE Trans. Comput. Aided Des. Integr. Circuits Syst. **25**(10), 2949–2961 (2017). https://doi.org/10.1109/TVLSI.2017.2717844

[Ake78] S.B. Akers, Binary desicion diagrams. IEEE Trans. Comput. 27 (1978)

[Alb05] C. Albrecht, IWLS 2005 benchmarks, in *Proceedings of the International Workshop on Logic & Synthesis* (2005)

[BA13] M. Bushnell, V. Agrawal, *Essentials of Electronic Testing for Digital, Memory and Mixed-Signal VLSI Circuits* (Springer, 2013). https://doi.org/10.1007/b117406

[Bau05] R.C. Baumann, Radiation-induced soft errors in advanced semiconductor technologies. IEEE Trans. Device Mater. Reliab. **5**(3), 305–316 (2005). https://doi.org/10.1109/TDMR.2005.853449

[BEI09] O. Ben-Kiki, C. Evans, B. Ingerson, YAML ain't markup language (yaml™) version 1.1, in *Working Draft 2008-05*, 11 (2009)

[Ben14] P. Bennett, *The Why, Where and What of Low-Power SoC Design*, 02/22/2020 (2014). https://www.eetimes.com/the-why-where-and-what-of-low-power-soc-design/#

[Bie+09] A. Biere et al., *Handbook of Satisfiability*, vol. 185. Frontiers in AI and Applications (IOS Press, 2009)

[Bie+99a] A. Biere et al., Symbolic model checking using SAT procedures instead of BDDs, in *Proceedings of the Design Automation Conference* (1999), pp. 317–320. https://doi.org/10.1109/DAC.1999.781333

[Bie+99b] A. Biere et al., Symbolic model checking without BDDs, in *Proceedings of the International Conference on Tools and Algorithms for the Construction and Analysis of Systems* (Springer, 1999), pp. 193–207. https://doi.org/10.1007/3540490590_14

[Bie08] A. Biere, Adaptive restart strategies for conflict driven SAT solvers, in *Proceedings of the International Conference on Theory and Application of Satisfiability Testing* (Springer, Berlin, Heidelberg, 2008), pp. 28–33 https://doi.org/10.1007/9783540797197_4

[Bla+08] D. Blaauw et al., Razor II: In situ error detection and correction for PVT and SER tolerance, in *Proceedings of the IEEE International Conference on Solid-State Circuits* (2008), pp. 400–622. https://doi.org/10.1109/ISSCC.2008.4523226

[Bry86] R.E. Bryant, Graph-based algorithms for Boolean function manipulation. IEEE Trans. Comput. Aided Des. Integr. Circuits Syst. **C-35**(8), 677–691 (1986). https://doi.org/10.1109/TC.1986.1676819

S. Huhn, R. Drechsler, *Design for Testability, Debug and Reliability*,
https://doi.org/10.1007/978-3-030-69209-4

[BS96] M. Böhm, E. Speckenmeyer, A fast parallel SAT-solver — efficient workload balancing. Ann. Math. Artif. Intell. **17**(2), 381–400 (1996). https://doi.org/10.1007/BF02127976

[BT07] K.J. Balakrishnan, N.A. Touba, Relationship between entropy and test data compression. IEEE Trans. VLSI Syst. **26**(2), 386–395 (2007). https://doi.org/10.1109/TCAD.2006.882600

[BW89] A.E. Barbour, A.S. Wojcik, A general constructive approach to fault-tolerant design using redundancy. IEEE Trans. Comput. **38**(1), 15–29 (1989). https://doi.org/10.1109/12.8727

[CC01] A. Chandra, K. Chakrabarty, System-on-a-chip test-data compression and decompression architectures based on Golomb codes. IEEE Trans. VLSI Syst. **20**(3), 355–368 (2001). https://doi.org/10.1109/43.913754

[Cla+18] E.M. Clarke et al., *Model Checking* (MIT press, 2018)

[Cla06] C.J. Clark, *CJTAG: Enhancement to IEEE 1149.1 Uses Concurrent Test to Reduce Test Times* (Kluwer Academic Publishers, 2006)

[Coo71] S.A. Cook, The complexity of theorem-proving procedures, in *Proceedings of the ACM International Symposium on the Theory of Computing* (Shaker Heights, Ohio, USA, 1971), pp. 151–158. https://doi.org/10.1145/800157.805047

[DAC99] M.B. Dwyer, G.S. Avrunin, J.C. Corbett, Patterns in property specifications for finite-state verification, in *Proceedings of the International Conference on Software Engineering* (1999), pp. 411–420. https://doi.org/10.1145/302405.302672

[DC14] S. Deutsch, K. Chakrabarty, Massive signal tracing using on-chip DRAM for in-system silicon debug, in *Proceedings of the International Test Conference* (2014), pp. 1–10. https://doi.org/10.1109/TEST.2014.7035363

[DDG00] R. Drechsler, N. Drechsler, W. Gunther, Fast exact minimization of BDD's. IEEE Trans. Comput. Aided Des. Integr. Circuits Syst. **19**(3), 384–389 (2000). https://doi.org/10.1109/43.833206

[Dho+16] H. Dhotre et al., Automated optimization of scan chain structure for test compression-based designs, in *Proceedings of the IEEE Asian Test Symposium* (2016), pp. 185–190. https://doi.org/10.1109/ATS.2016.60

[DLL62] M. Davis, G. Logemann, D. Loveland, A machine program for theorem-proving. Commun. ACM **5**(7), 394–397 (1962). https://doi.org/10.1145/368273.368557

[DM13] W.R.A. Dias, E.D. Moreno, Code compression using multi-level dictionary, in *IEEE 4th Latin American Symposium on Circuits and Systems (LASCAS)* (2013), pp. 1–4. https://doi.org/10.1109/LASCAS.2013.6519043

[Doy+10] L. Doyen et al., Robustness of sequential circuits, in *Proceedings of the International Conference on Application of Concurrency to System Design* (2010), pp. 77–84. https://doi.org/10.1109/ACSD.2010.26

[DP60] M.D. Davis, H. Putnam, A computing procedure for quantification theory. J. Assoc. Comput. Mach. **7**, 201–215 (1960)

[Dre+09] R. Drechsler et al., *Test Pattern Generation Using Boolean Proof Engines* (Springer, 2009). https://doi.org/10.1007/978-90-481-2360-5

[DS07] S. Disch, C. Scholl, Combinational equivalence checking using incremental SAT solving, output ordering, and resets, in *Proceedings of the ASP Design Automation Conference* (2007), pp. 938–943. https://doi.org/10.1109/ASPDAC.2007.358110

[DS91] B.I. Dervisoglu, G.E. Stong, Design for testability using scanpath techniques for path-delay test and measurement, in *Proceedings of the International Test Conference* (1991), pp. 365–374. https://doi.org/10.1109/TEST.1991.519696

[DSF04] R. Drechsler, J. Shi, G. Fey, Synthesis of fully testable circuits from BDDs. IEEE Trans. Comput. Aided Des. Integr. Circuits Syst. **23**(3), 440–443 (2004). https://doi.org/10.1109/TCAD.2004.823342

[ED10] S. Eggersglüß, R. Drechsler, Robust algorithms for high quality test pattern generation using Boolean satisfiability, in *Proceedings of the International Test Conference* (2010), pp. 1–10. https://doi.org/10.1109/TEST.2010.5699289

[Egg+16a] S. Eggersglüß et al., Formal test point insertion for region-based low-capture-power compact at-speed scan test, in *Proceedings of the IEEE Asian Test Symposium* (2016), pp. 173–178. https://doi.org/10.1109/ATS.2016.41

[Egg+16b] S. Eggersglüß et al., On optimization-based ATPG and its application for highly compacted test sets. IEEE Trans. Comput. Aided Des. Integr. Circuits Syst. **35**(12), 2104–2117 (2016) https://doi.org/10.1109/TCAD.2016.2552822

[Eld59] R.D. Eldred, Test routines based on symbolic logical statements. J. ACM **6**(1), 33–37 (1959). https://doi.org/10.1145/320954.320957

[EMW16] S. Eggersglüß, K. Miyase, X. Wen, SAT-based post-processing for regional capture power reduction in at-speed scan test generation, in *Proceedings of the IEEE European Test Symposium* (2016), pp. 1–6. https://doi.org/10.1109/ETS.2016.7519327

[Ern+03] D. Ernst et al., Razor: a low-power pipeline based on circuit-level timing speculation, in *Proceedings of the IEEE/ACM International Symposium on Microarchitecture* (2003), pp. 7–18. https://doi.org/10.1109/MICRO.2003.1253179

[ES04] N. Eén, N. Sörensson, An extensible SAT-solver, in *Lecture Notes in Computer Science 2919* (2004), pp. 502–518. https://doi.org/10.1007/978-3-540-24605-3_37

[EW77] E.B. Eichelberger, T.W. Williams, A logic design structure for lsi testability, in *Proceedings of the Design Automation Conference* (IEEE Press, 1977), pp. 462–468. https://doi.org/10.1145/62882.62924

[EWD13] S. Eggersglüß, R. Wille, R. Drechsler, Improved SAT-based ATPG: More constraints, better compaction, in *Proceedings of the International Conference on Computer-Aided Design* (2013), pp. 85–90. https://doi.org/10.1109/ICCAD.2013.6691102

[FD08] G. Fey, R. Drechsler, A basis for formal robustness checking, in *Proceedings of the International Symposium on Quality Electronic Design* (2008), pp. 784–789. https://doi.org/10.1109/ISQED.2008.4479838

[Fey+11] G. Fey et al., Effective robustness analysis using bounded model checking techniques. IEEE Trans. Comput. Aided Des. Integr. Circuits Syst. **30**(8), 1239–1252 (2011) https://doi.org/10.1109/TCAD.2011.2120950

[FFM08] N. Farazmand, M. Fazeli, S.G. Miremadi, FEDC: Control flow error detection and correction for embedded systems without program interruption, in *Third International Conference on Availability, Reliability and Security* (2008), pp. 33–38. https://doi.org/10.1109/ARES.2008.199

[FS83] H. Fujiwara, T. Shimono, On the acceleration of test generation algorithms. IEEE Trans. Comput. **C-32**(12), 1137–1144 (1983). https://doi.org/10.1109/TC.1983.1676174

[FSF14] A. Finder, A. Sülflow, G. Fey, Latency analysis for sequential circuits. IEEE Trans. Comput. Aided Des. Integr. Circuits Syst. **33**(4), 643–647 (2014). https://doi.org/10.1109/TCAD.2013.2292501

[GC13] S.K. Goel, K. Chakrabarty, *Testing for Small-Delay Defects in Nanoscale CMOS Integrated Circuits* (CRC Press, 2013). https://doi.org/10.1201/b15549

[Geb+07] M. Gebser et al., Conflict-driven answer set solving, in *Proceedings of the International Joint Conference on Artificial Intelligence* (2007), pp. 386–392. https://doi.org/10.1016/j.artint.2012.04.001

[GMG90] M. Gerner, B. Müller, G. Sandweg, *Selbsttest digitaler Schaltungen* (Springer, 1990)

[GN02] E. Goldberg, Y. Novikov, BerkMin: A fast and robust SAT-solver, in *Proceedings of the Design, Automation and Test in Europe* (2002), pp. 142–149. https://doi.org/10.1109/DATE.2002.998262

[GNR64] J.M. Galey, R.E. Norby, J.P. Roth, Techniques for the diagnosis of switching circuit failures. IEEE Trans. Commun. Electron. **83**(74), 509–514 (1964) https://doi.org/10.1109/TCOME.1964.6539498

[GR81] P. Goel, B.C. Rosales, PODEM-X: An automatic test generation system for VLSI logic structures, in *Proceedings of the Design Automation Conference* (1981), pp. 260–268. https://doi.org/10.1109/DAC.1981.1585361

[Gru+05] O. Grumberg et al., Achieving speedups in distributed symbolic reachability analysis through asynchronous computation, in *Proceedings of the Advanced Research Working Conference on Correct Hardware Design and Verification Methods* (Springer, 2005), pp. 129–145. https://doi.org/10.1007/11560548_12

[Gru09] O. Grumberg, 3-valued abstraction for (bounded) model checking, in *Proceedings of the International Conference Automated Technology for Verification and Analysis* (Springer, 2009), pp. 21–21. https://doi.org/10.1007/978-3-642-04761-9_2

[Gut+01] M.R. Guthaus et al., MiBench: A free, commercially representative embedded benchmark suite, in *Proceedings of the IEEE International Workshop on Workload Characterization* (2001), pp. 3–14. https://doi.org/10.1109/WWC.2001.99739

[Ham50] R.W. Hamming, Error detecting and error correcting codes. Bell Syst. Tech. J. **29**(2), 147–160 (1950). https://doi.org/10.1002/j.1538-7305.1950.tb00463.x

[HED16a] S. Huhn, S. Eggersglüß, R. Drechsler, Leichtgewichtige Datenkompressions-Architektur für IEEE-1149.1-kompatible Testschnittstellen, in *Informal Proceedings of the GI/GMM/ITG Workshop für Testmethoden und Zuverlässigkeit von Schaltungen und Systemen* (2016)

[HED16b] S. Huhn, S. Eggersglüß, R. Drechsler, VecTHOR: Low-cost compression architecture for IEEE-1149.1-compliant TAP controllers, in *Proceedings of the IEEE European Test Symposium* (2016), pp. 1–6. https://doi.org/10.1109/ETS.2016.7519303

[HED17] S. Huhn, S. Eggersglüß, R. Drechsler, Reconfigurable TAP controllers with embedded compression for large test data volume, in *Proceedings of the IEEE Defect and Fault Tolerance in VLSI and Nanotechnology Systems* (2017), pp. 1–6. https://doi.org/10.1109/DFT.2017.8244462

[HED19] S. Huhn, S. Eggersglüß, R. Drechsler, *Enhanced Embedded Test Compression Technique For Processing Incompressible Test Patterns*. Informal Proceedings of the GI/GMM/ITG Workshop für Testmethoden und Zuverlässigkeit von Schaltungen und Systemen (2019)

[Het+99] G. Hetherington et al., Logic BIST for large industrial designs: real issues and case studies, in *Proceedings of the International Test Conference* (1999), pp. 358–367. https://doi.org/10.1109/TEST.1999.805650

[HN06] T. Heijmen, A. Nieuwland, Soft-error rate testing of deep-submicron integrated circuits, in *Proceedings of the IEEE European Test Symposium* (2006), pp. 247–252. https://doi.org/10.1109/ETS.2006.42

[HSV13] S. Hertz, D. Sheridan, S. Vasudevan, Mining hardware assertions with guidance from static analysis. IEEE Trans. Comput. Aided Des. Integr. Circuits Syst. **32**(6), 952–965 (2013). https://doi.org/10.1109/TCAD.2013.2241176

[HTD19a] S. Huhn, D. Tille, R. Drechsler, A hybrid embedded multi-channel test compression architecture for low-pin count test environments in safety-critical systems, in *Proceedings of the International Test Conference in Asia* (2019), pp. 115–120. https://doi.org/10.1109/ITC-Asia.2019.00033

[HTD19b] S. Huhn, D. Tille, R. Drechsler, Hybrid architecture for embedded test compression to process rejected test patterns, in *Proceedings of the IEEE European Test Symposium* (2019), pp. 1–2. https://doi.org/10.1109/ETS.2019.8791508

[Hua+98] S.-Y. Huang et al., Fault-simulation based design error diagnosis for sequential circuits, in *Proceedings of the Design Automation Conference* (1998), pp. 632–637. https://doi.org/10.1109/DAC.1998.724548

[Huh+17a] S. Huhn et al., Enhancing robustness of sequential circuits using application-specific knowledge and formal methods, in *Proceedings of the Asia and South Pacific Design Automation Conference* (2017), pp. 182–187. https://doi.org/10.1109/ASPDAC.2017.7858317

[Huh+17b] S. Huhn et al., Optimization of retargeting for IEEE 1149.1 TAP controllers with embedded compression, in *Proceedings of the IEEE Design, Automation and Test in Europe* (2017), pp. 578–583. https://doi.org/10.23919/DATE.2017.7927053

[Huh+18] S. Huhn et al., *A Codeword-Based Compaction Technique for On-Chip Generated Debug Data Using Two-Stage Artificial Neural Ntworks*. Informal Proceedings of the GI/GMM/ITG Workshop für Testmethoden und Zuverlässigkeit von Schaltungen und Systemen (2018)

[Huh+19] S. Huhn et al., Determing application-specific knowledge for improving robustness of sequential circuits. IEEE Trans. Very Large Scale Integr. Syst., 875–887 (2019). https://doi.org/10.1109/TVLSI.2018.2890601

[01] IEEE Standard Verilog Hardware Description Language, in *IEEE Std 1364-2001* (2001), pp. 1–792. https://doi.org/10.1109/IEEESTD.2001.93352

[13] IEEE Standard for Test Access Port and Boundary-Scan Architecture Redline, in *IEEE Std 1149.1-2013 (Revision of IEEE Std 1149.1-2001) Redline* (2013), pp. 1–899

[14] IEEE Standard for Access and Control of Instrumentation Embedded Within a Semiconductor Device, in *IEEE Std 1687-2014* (2014), pp. 1–283

[Cor11] Intel Corporation, *The Story of the Intel 4004 - Intel's First Microprocessor*, 02/22/2020 (2011). https://www.intel.de/content/www/de/de/history/museum-story-of-intel-4004.html

[Cor13] Intel Corporation, *LGA1150 Socket - Application Guide*, 02/21/2020 (2013). https://www.intel.com/content/dam/www/public/us/en/documents/guides/4th-gen-core-lga1150-socket-guide.pdf

[ICM99] V. Iyengar, K. Chakrabarty, B.T. Murray, Deterministic built-in pattern generation for sequential circuits. J. Electron. Test. **15**(1-2), 97–114 (1999). https://doi.org/10.1023/A:1008384201996

[Ila+13] K. Ilambharathi et al., Domain Specific Hierarchical Huffman Encoding, in *Cornell University Library*, abs/1307.0920 (2013)

[Inc20] Verific Design Automation Inc., *Verific's Parser Platform*, 02/15/2020 (2020). https://www.verific.com

[Iye+02] V. Iyengar et al., Test resource optimization for multi-site testing of SOCs under ATE memory depth constraints, in *Proceedings of the International Test Conference* (2002), pp. 1159–1168. https://doi.org/10.1109/TEST.2002.1041874

[Jan+11] J. Janicki et al., EDT channel bandwidth management in SoC designs with pattern-independent test access mechanism, in *Proceedings of the International Test Conference* (2011), pp. 1–9. https://doi.org/10.1109/TEST.2011.6139170

[JB18] S. Jayanthy, M.C. Bhuvaneswari, *Test Generation of Crosstalk Delay Faults in VLSI Circuits* (Springer, 2018). https://doi.org/10.1007/978-981-13-2493-2

[JG02] N.K. Jha, S. Gupta, *Testing of Digital Systems* (Cambridge University Press, USA, 2002). https://doi.org/10.1017/CBO9780511816321

[JGT99] A. Jas, J. Ghosh-Dastidar, N.A. Touba, Scan vector compression/decompression using statistical coding, in *Proceedings of the VLSI Test Symposium* (1999), pp. 114–120. https://doi.org/10.1109/VTEST.1999.766654

[Jia+12] R. Jia et al., JTAG-based bitstream compression for FPGA configuration, in *Proceedings of the International Conference on Solid-State and Integrated Circuit Technology* (2012), pp. 1–3. https://doi.org/10.1109/ICSICT.2012.6467807

[JT98] A. Jas, N.A. Touba, Test vector decompression via cyclical scan chains and its application to testing core-based designs, in *Proceedings of the International Test Conference* (1998), pp. 458–464. https://doi.org/10.1109/TEST.1998.743186

[JW90] R.G. Jeroslow, J. Wang, Solving propositional satisfiability problems. Ann. Math. Artif. Intell. **1**(1), pp. 167–187 (1990). https://doi.org/10.1007/BF01531077

[Kar05] B. Karlsson, *Beyond the C++ Standard Library: An Introduction to Boost* (Pearson Education, 2005)

[KCR06] F.L. Kastensmidt, L. Carro, R. Reis, *Fault-Tolerance Techniques for SRAM-Based FPGAs* (Springer, 2006). https://doi.org/10.1007/978-0-387-31069-5

[KL93] H. Konuk, T. Larrabee, Explorations of sequential ATPG using Boolean satisfiability, in *Proceedings of the VLSI Test Symposium* (1993), pp. 85–90. https://doi.org/10.1109/VTEST.1993.313303

[Kra+06] U. Krautz et al., Evaluating coverage of error detection logic for soft errors using formal methods, in *Proceedings of the Design, Automation and Test in Europe* (2006), pp. 1–6. https://doi.org/10.1109/DATE.2006.244062

[KS03] D. Kececioglu, F.-B. Sun, *Environmental Stress Screening: Its Quantification, Optimization and Management* (DEStech Publications, Inc, 2003)

[LA04] C. Lattner, V. Adve, LLVM: a compilation framework for lifelong program analysis transformation, in *Proceedings of the International Symposium on Code Generation and Optimization* (2004), pp. 75–86. https://doi.org/10.1109/CGO.2004.1281665

[Lar11] R.S. Laramee, *Bob's Concise Introduction to Doxygen*. Tech. rep. Technical report, The Visual and Interactive Computing Group, Computer Science Department (2011). https://cs.swan.ac.uk/~csbob/teaching/laramee07commentConvention.pdf

[Lar92] T. Larrabee, Test pattern generation using Boolean satisfiability. IEEE Trans. Comput. Aided Des. Integr. Circuits Syst. **11**(1), 4–15 (1992). https://doi.org/10.1109/43.108614

[LC03] L. Li, K. Chakrabarty, Test data compression using dictionaries with fixed-length indices, in *Proceedings of the VLSI Test Symposium* (2003), pp. 219–224. https://doi.org/10.1109/VTEST.2003.1197654

[Lee59] C.Y. Lee, Representation of switching circuits by binary-decision programs. Bell Syst. Tech. J. **38**(4), 985–999 (1959). https://doi.org/10.1002/j.1538-7305.1959.tb01585.x

[Lie06] J. Lienig, *Layoutsynthese elektronischer Schaltungen Grundlegende Algorithmen für die Entwurfsautomatisierung* (Springer, 2006). https://doi.org/10.1007/3-540-29942-4

[Loi+13] C. Loiacono et al., Fast Cone-Of-Influence computation and estimation in problems with multiple properties, in *Proceedings of the Design, Automation and Test in Europe* (2013), pp. 803–806. https://doi.org/10.7873/DATE.2013.170

[LR12] X. Lin, J. Rajski, On utilizing test cube properties to reduce test data volume further, in *Proceedings of the IEEE Asian Test Symposium* (2012), pp. 83–88. https://doi.org/10.1109/ATS.2012.41

[Luo89] Z. Luo, ECC, an extended calculus of constructions, in *Proceedings of the International Symposium on Logic in Computer Science* (1989), pp. 386–395. https://doi.org/10.1109/LICS.1989.39193

[McM03] K.L. McMillan, Interpolation and SAT-based model checking, in *Proceedings of the International Conference on Computer-Aided Verification* (Springer, 2003), pp. 1–13. https://doi.org/10.1007/978-3-540-45069-6_1

[Mic03] A. Miczo, *Digital Logic Testing and Simulation*, Vol. 2 (Wiley, 2003)

[MK06] S. Mitra, K.S. Kim, XPAND: an efficient test stimulus compression technique. IEEE Trans. Comput. **55**(2), 163–173 (2006). https://doi.org/10.1109/TC.2006.31

[MM10] N. Miskov-Zivanov, D. Marculescu, Multiple transient faults in combinational and sequential circuits: A systematic approach. IEEE Trans. Comput. Aided Des. Integr. Circuits Syst. **29**(10), 1614–1627 (2010). https://doi.org/10.1109/TCAD.2010.261131

[Moh09] I. Mohor, *JTAG Test Access Port (TAP)*, 02/15/2020 (2009). http://opencores.org/projects/jtag

[Moo65] G.E. Moore, Cramming more components onto integrated circuits. Electronics **38**(8), 539–535 (1965)

[Mos+01] M.W. Moskewicz et al., Chaff: engineering an efficient SAT solver, in *Proceedings of the Design Automation Conference* (2001), pp. 530–535. https://doi.org/10.1145/378239.379017

[MR16] N. Mukherjee, J. Rajski, Digital testing of ICs for automotive applications, in *Proceedings of the International Conference on VLSI Design* (2016), pp. 14–16. https://doi.org/10.1109/VLSID.2016.134

[MS00] J. Marques-Silva, K. Sakallah, Invited tutorial: Boolean satisfiability algorithms and applications in electronic design automation, in *Proceedings of the International Conference on Computer-Aided Verification* (Springer, Berlin, Heidelberg, 2000), pp. 3–3. https://doi.org/10.1007/10722167_3

[MS99] J.P. Marques-Silva, K.A. Sakallah, GRASP: A search algorithm for propositional satisfiability. IEEE Trans. Comput. **48**(5), 506–521 (1999). https://doi.org/10.1109/12.769433

[Muj19] H. Mujtaba, *AMD 2nd Gen EPYC Rome Processors Feature A Gargantuan 39.54 Billion Transistors, IO Die Pictured in Detail*, 02/15/2020 (2019). https://wccftech.com/amd-2nd-gen-epyc-rome-iod-ccd-chipshots-39-billion-transistors/

[MV08] T.M. Mak, S. Venkataraman, Design for debug and diagnosis, in *System-on-Chip Test Architectures* (2008), pp. 463–504. https://doi.org/10.1016/B978-012373973-5.50015-2

[Nag14] P. Nagaraj, *Test Cost Challenges in LPCT (Low Pin Count Test) Designs*, 02/22/2020 (2014). https://www.edn.com/test-cost-challenges-in-lpct-low-pin-count-test-designs/

[NW99] D. Nayak, D.M.H. Walker, Simulation-based design error diagnosis and correction in combinational digital circuits, in *Proceedings of the VLSI Test Symposium* (1999), pp. 70–78. https://doi.org/10.1109/VTEST.1999.766649

[Raj+04] J. Rajski et al., Embedded deterministic test. IEEE Trans. VLSI Syst. **23**(5), 776–792 (2004). https://doi.org/10.1109/TCAD.2004.826558

[Rie+16] H. Riener et al., metaSMT: focus on your application and not on solver integration. Int. J. Softw. Tool Technol. Trans., 1–17 (2016)

[Rot66] J.P. Roth, Diagnosis of automata failures: A calculus and a method IBM J. Res. Dev. **10**(4), 278–291 (1966). https://doi.org/10.1147/rd.104.0278

[RR02] S.K. Roy, S. Ramesh, Functional verification of system on chips practices, issues and challenges, in *Proceedings of the ASP Design Automation Conference* (2002), pp. 11–13. https://doi.org/10.1109/ASPDAC.2002.994873

[RS95] K. Ravi, F. Somenzi, High-density reachability analysis, in *Proceedings of the International Conference on Computer-Aided Design* (1995), pp. 154–158. https://doi.org/10.1109/ICCAD.1995.480006

[SB89] C.E. Stroud, A.E. Barbour, Design for testability and test generation for static redundancy system level fault-tolerant circuits, in *Proceedings of the International Test Conference* (1989), pp. 812–818. https://doi.org/10.1109/TEST.1989.82370

[SBP16] M. Sauer, B. Becker, I. Polian, PHAETON: A SAT-based framework for timing-aware path sensitization. IEEE Trans. Comput. **65**(6), 1869–1881 (2016). https://doi.org/10.1109/TC.2015.2458869

[SC85] A.P. Sistla, E.M. Clarke, The complexity of propositional linear temporal logics. J. ACM **32**(3), 733–749 (1985). https://doi.org/10.1145/3828.3837

[Sch+99] C. Scholl et al., BDD minimization using symmetries. IEEE Trans. Comput. Aided Des. Integr. Circuits Syst. **18**(2), 81–100 (1999). https://doi.org/10.1109/43.743706

[Som16] F. Somenzi, *CUDD: CU Decision Diagram Package* (University of Colorado at Boulder, 2016). ftp://vlsi.colorado.edu/pub/. 2016-04-11

[SP93] J. Savir, S. Patil, Scan-based transition test. IEEE Trans. Comput. Aided Des. Integr. Circuits Syst. **12**(8), 1232–1241 (1993). https://doi.org/10.1109/43.238615

[SP94] J. Savir, S. Patil, Broad-side delay test. IEEE Trans. Comput. Aided Des. Integr. Circuits Syst. **13**(8), 1057–1064 (1994). https://doi.org/10.1109/43.298042

[SRM14] A. Sanchez-Macian, P. Reviriego, J.A. Maestro, Hamming SEC-DAED and extended hamming SEC-DED-TAED codes through selective shortening and bit placement. IEEE Trans. Device Mater. Reliab. **14**(1), 574–576 (2014). https://doi.org/10.1109/TDMR.2012.224753

[TD09] D. Tille, R. Drechsler, A fast untestability proof for SAT-based ATPG, in *Proceedings of the IEEE Symposium on Design and Diag. of Electronic Circuits and Systems* (2009), pp. 38–43. https://doi.org/10.1109/DDECS.2009.5012096

[TED10] D. Tille, S. Eggersgluss, R. Drechsler, Incremental solving techniques for SAT-based ATPG. IEEE Trans. Comput. Aided Des. Integr. Circuits Syst. **29**(7), 1125–1130 (2010). https://doi.org/10.1109/TCAD.2010.2044673

[Tse83] G. Tseitin, On the complexity of derivation in propositional calculus, in *Studies in Constructive Mathematics and Mathematical Logic, Part 2*, Vol. 8 (Springer, 1983), pp. 466–483. https://doi.org/10.1007/978-3-642-81955-1_28

[Vel04] M.N. Velev, Encoding global unobservability for efficient translation to SAT, in *Proceedings of the International Conference on Theory and Applications of Satisfiability Testing* (2004), pp. 197–204

[VG02] B. Vermeulen, S.K. Goel, Design for debug: catching design errors in digital chips. IEEE Des. Test Comput. **19**(3), 35–43 (2002). https://doi.org/10.1109/MDT.2002.103792

[VG14] B. Vermeulen, K. Goossens, The complexity of debugging system chips. *Debugging Systems-on-Chip: Embedded Systems* (Springer, 2014), pp. 139–155. https://doi.org/10.1007/978-3-319-06242-6_6

[Vic+05] D.W. Victor et al., Functional verification of the POWER5 microprocessor and POWER5 multiprocessor systems. IBM J. Res. Dev. **49**(4-5), 541–553 (2005). https://doi.org/10.1147/rd.494.0541

[WLR19] C. Wu, K. Lee, S.M. Reddy, An efficient diagnosis-aware ATPG procedure to enhance diagnosis resolution and test compaction. IEEE Trans. VLSI Syst. **27**(9), 2105–2118 (2019). https://doi.org/10.1109/TVLSI.2019.2919233

[WP02] F.G. Wolff, C. Papachristou, Multiscan-based test compression and hardware decompression using LZ77, in *Proceedings of the International Test Conference* (2002), pp. 331–339. https://doi.org/10.1109/TEST.2002.1041776

[WTH04] A. Wurtenberger, C.S. Tautermann, S. Hellebrand, Data compression for multiple scan chains using dictionaries with corrections, in *Proceedings of the International Test Conference* (2004), pp. 926–935. https://doi.org/10.1109/TEST.2004.1387357

[WW03] M. Weber, J. Weisbrod, Requirements engineering in automotive development-experiences and challenges. Trans. IEEE Softw. **20**(1), 16–24 (2003)

[WWW06] L.-T. Wang, C.-W. Wu, X. Wen, *VLSI Test Principles and Architectures: Design for Testability (Systems on Silicon)* (Morgan Kaufmann Publishers, San Francisco, CA, USA, 2006)

[ZL77] J. Ziv, A. Lempel, A universal algorithm for sequential data compression. IEEE Trans. Inf. Theory **23**(3), 337–343 (1977). https://doi.org/10.1109/TIT.1977.1055714

[ZS03] Y. Zorian, S. Shoukourian, Embedded-memory test and repair: infrastructure IP for SoC yield. IEEE Trans. Design Test Comput. **20**(3), 58–66 (2003). https://doi.org/10.1109/MDT.2003.1198687

Index

S. Huhn, R. Drechsler, *Design for Testability, Debug and Reliability*,
https://doi.org/10.1007/978-3-030-69209-4

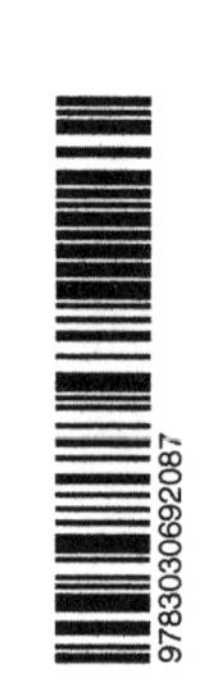